# Tanks & Terminals

A 'primer' for Engineers
new to the bulk liquid
storage sector

A.R.Polain

# Contents

# TERMINAL ENGINEERING

# LOSS PREVENTION

# TERMINAL TYPES

* * *

# Appendices

· · · · · ● ● ● ● ● ● ● ● ● ● ● ● · · · · ·

# *Preface*

By modern standards, the dictionary definition of '**Engineer**' is woefully inadequate.

The description of an Engineer as *"a person whose job involves designing and building engines, machines, roads, bridges, etc."* may have been valid in the 18th Century. However, by the 21st Century, Engineering had evolved into a multitude of separate disciplines, each containing a variety of specializations, few of which have anything to do with engines or bridges.

The early Engineers did not just have the foresight to identify problems and the ingenuity to solve those problems. They had a background in crafts and trades and respect for the artisans who converted their revolutionary ideas to practical reality.

As the years have passed, and access to university has become easier and more common, practical experience of a craft or trade is considered less important than academic study.

Not that long ago, thick and thin engineering sandwich degree courses were popular. These were degrees that included a period of on-the-job training as part of the course requirement. During the industrial placement, undergraduates would be exposed to the realities of life in an engineering environment and gain some understanding of the breadth and variety of skills and talents needed. Unfortunately, universities seem less inclined to run sandwich courses these days.

In the past, many of those who did a straight 3 years at University were employed by large engineering and manufacturing companies with post graduation training courses. Spending 2 years rotating between departments was not unusual. This frequently involved a period of workshop training or assignment to a construction site. However, with the decline of the engineering and manufacturing sector in the UK (and elsewhere) few graduates these days have an opportunity to gain any practical knowledge before starting their career in earnest.

A stock response to these complaints is that undergraduates learn it all at university. But what chance do they have when they are being taught by a generation of lecturers and professors who have no practical experience themselves?

Another problem is that in the 'old days' it was easy to bounce around from one branch of engineering to another, picking up experience as you went. Opportunities existed for working in a design office, then moving to the construction site (although opportunities to move the other way have always been more limited). There were more companies, more jobs, and more opportunities for those willing to search them out.

Nowadays, graduates tend to move into an industry and stay in that industry. They are expected to know it all, and, at least initially, they believe it themselves.

I first saw this problem when a young graduate sheepishly approached me to say that although his job involved supervising welders he did not know anything about welding (they do not teach you welding at University). In that case, the problem was easy to solve. I spoke to my contractors and arranged for all the graduates to spend a few days working alongside the welders and pipefitters. By all accounts, it was very successful and enjoyed especially by the welders, but it did raise the question about what else the graduates needed in the way of practical experience.

I do not suppose the situation is any better these days. At the Engineering Consultancy I last worked for, they took on young graduates and immediately stuck them in management and supervisory roles. The company convinced itself that the graduates would gain experience by some sort of magical osmosis by sitting near more experienced people-anything as long as it did not cost the Consultancy any money to train people. Strangely, Clients seemed amazingly tolerant of paying top dollar for 'Consultants' when all they were getting were inexperienced and untrained 'newbies', learning at the Client's expense and in the Client's time.

When I retired, it seemed a shame that the knowledge and experience I had accumulated in almost 50 years would disappear with me. Therefore, I decided the least I could do was to make available what I knew to the next generation (if anyone was remotely interested!).

My first effort dealt with **'Financial Transactions'** – *an Engineers guide to an Engineers role*.

What started as a few weeks in the winter months ended up taking a few years. But I learned some useful lessons, and was encouraged to repeat the process with a new volume "**Tanks and Terminals**".

I do not claim to be an 'world expert' on **Tanks and Terminals** (or anything else for that matter). I am not even sure who qualifies as an expert these days if a 15-year-old Swedish schoolgirl can be the world's foremost expert on climate change.

Over the years I estimate I have managed the construction of 100 plus new tanks, been the maintenance supervisor in a large tank farm, and have audited 70 or 80 Terminals of various types. I have advised on Terminal design, supervised construction, and even knocked a few Terminals down (on purpose of course). Working as a Technical Advisor for Banks and Investors, I oversaw plans for 'green field' developments, Terminal extensions, revamps, debottlenecking and management restructuring.

In the process, I hope I have picked up some information that others may benefit from. If that is so, the time spent writing this book will not have been wasted.

·····•●•●•●•●•●•●●•●•·····

# Introduction

This is **NOT** a textbook. It is a '**primer**' – defined as a small introductory book on a subject.

There are plenty of proper textbooks on the subject of **Tanks** and even a few that deal with **Terminals**, but who has the time or inclination to plough through a weighty tome? So often, these books are written by academics that went straight from 'undergraduate' to 'lecturer' without the inconvenience of actually working in industry for a living.

Have you seen how many books there are about construction, written by professors? When was the last time you heard of a professor working on a construction site, or a construction engineer becoming a professor? Me neither.

In this 'primer', I have approached **Tanks and Terminals** from a practical perspective.

I have written about what I know firsthand, not what you may see in a reference book. To get the detail correct I have resorted to published sources (Google is my constant companion and should get a credit somewhere) but basically, this is the distilled essence of everything I have picked up at site or working with Terminal Owners and Operators.

I have broken this work into self-contained sections, initially looking at the life cycle of tanks, followed by the topics that can influence the operation of a Terminal. Conscious that this is a 'primer', I have **highlighted** the key standards and specifications so the reader can look up the topics in more detail if wanted. Over the years, I have accumulated various publications on tanks and terminals (many Engineers are secret squirrels), and where these are now available for free on the internet, I have referred to them.

***QUESTION:   Is this book exhaustive, authoritative and definitive?***

***ANSWER:   No!***

There are too many Terminals, constructed in different ways, operated in different ways, and built for different purposes for any single book to cover every conceivable topic in depth. Every Terminal will have or do something unique and every Terminal Owner and Operator will be convinced that the way they are doing things is the best possible way and everyone else is wrong.

Nevertheless, there is seldom a 'right' or 'wrong' way of doing things. To quote an old saying "there is more than one way to skin a cat" and there is certainly more than one way to build and run a Terminal.

I have specifically avoided some interesting topics and only briefly mentioned others (LNG Terminals) as these are not representative of the bulk liquid storage industry as a whole. If you want to find out more about these topics, there are plenty of resources online.

I have tried to touch on the topics that will be common in 80% of the Terminals you may visit. Moreover, I am a firm believer in site visits. You can learn more in one hour on-site than you can by reading a dozen books. I am also a believer in asking questions. You will be amazed at how happy most Terminal Operators are to share their knowledge and experience with outsiders. So, if you have the opportunity to visit a Terminal, ask many questions and listen closely to the Operator's response.

So who was this 'primer' written for?

I have presumed a certain amount of pre-existing engineering knowledge, so would recommend this book to any engineering student wanting to know more about the bulk liquid storage industry. It may also be of help to newly qualified graduates who wish to expand their knowledge about tanks or terminals. Operators in terminals may find the contents of some help, explaining topics they have taken for granted without knowing the background or logic. I think most of us have struggled with ISO 9000 without realizing how it has changed over the years or its actual distant relationship with 'quality'.

And then there is the qualified and experienced Engineer who has never been involved with tanks or terminals but needs to 'gen up' for a specific reason. Maybe they have been asked to do a 'Due Diligence' audit on a Terminal (in this case they should also be reading '**Financial Transactions**-*an Engineers guide to an Engineers role*' as well, as it includes some useful resources!).

To avoid any confusion with terminology, I have tried to be consistent.

- A **Terminal Owner** is a company or individual that owns and operates the Terminal;
- The **Terminal Operators** are the boys and girls who make the Terminal work. They swing the valves and push the buttons 365/7/24;
- Occasionally I may refer to a small Terminal as a Depot, but big or small, it is still a Terminal;
- In the case of audits, the **Engineer** is the person in charge of the audit. Frequently there will be specialists or junior staff who assist. The **Employer** is the organisation that employs the Engineer or benefits from the audit.

And remember, don't get fixated. Terminals come in all different shapes and sizes! From the small, simple, isolated and remote ground fuel depots in Africa to the giant multi-product terminals like Koole (Botlek) in Rotterdam.

·········●●●●●●●●●·●········

# CHAPTER 1

# *Above Ground Storage Tanks*

## 1.1    Tank Configurations

There are many ways to classify a tank and no universally accepted method. A classification commonly employed is the internal pressure of the tank:

- **Atmospheric tanks**-by far the most common type of tank. Although called atmospheric, these tanks are frequently operated up to 0.03barg (0.5 psig) above atmospheric pressure. Generally abbreviated to AST (atmospheric storage tank);

- **Low pressure tanks**-are designed to operate from atmospheric pressure up to 1 atm or 15 psig. Low pressure tanks have design standards, such as API 620;

The vapour pressure of the liquid being stored, its flash point, and the tank's internal design pressure will influence the type of tank base, shell, and roof the Designer selects.

### Terminology

Like all businesses, Terminal employees use shorthand when talking to each other. These can be different in the various sectors of the industry and may even be used differently by Operators on the same Terminal (and frequently on the same shift!). In practice, they can mean whatever the speaker wants them to mean (as Humpty Dumpty said to Alice). The trick is not to take it too literally and see the word in context.

**Capacity** usually refers to the **rated capacity** of a tank rather than the **absolute capacity**. The HSE's "Safety and Environmental standards" required a whole Appendix to define the difference.

**Rated Capacity**(safe or pumpable capacity) therefore refers to the volume of the tank between the *Full* and the *Empty* positions. **Absolute Capacity** (nominal capacity) refers to the total volume that can be stored in the tank without reference to the Full and Empty positions.

**Deadstock** refers to the liquid remaining at the bottom of the tank when the tank is nominally *Empty*. To remove the deadstock from the tank, small-bore drain lines and drain pumps (sometimes called 'stripping' pumps) are used. Deadstock is usually only removed when the tank is being inspected or a different product will be stored in the tank. Also known as the tank 'heel' or tank bottom.

However, the word **Ullage** is the one that is misused most frequently. Strictly speaking, it is defined as "*the amount by which a container falls short of being full*" i.e. the difference between the current liquid level and the *Full* position of the tank.

Sometimes it is easier to call this 'headspace' to avoid any potential confusion.

## Fixed Roof

These types of tanks are good for low-volatility materials. For volatile materials, fixed roof tanks may require inert gas blanketing (usually nitrogen) in the space between the liquid and the tank roof to dilute the vapourbelow the Lower Explosive Limit (LEL).

As a rule of thumb, fixed-roof tanks are used to store liquids with true vapor pressures (TVP) of less than 10 kPa (absolute).Fixed roof tanks need a breather valve to prevent tank damage due to overpressure during liquid filling and vacuum when emptying.

- **Flat Roof**-only for very small diameter tanks with no appreciable internal pressure;

- **Cone Roof**-The roof is self-supporting and made in the form of a shallow inverted cone. These tanks are economical to build and are suitable for a wide variety of liquid products with low vapour pressure. It may have a central support column, as well as supporting steelwork;

- **Dome Roof** -similar to Cone Roof except the roof is more nearly approximate to a sphere;

- **Geodesic Dome roof**-an economically attractive alternative to cone roof tanks. They offer superior corrosion resistance and are self-supporting across their entire span, requiring no internal support structure or supporting columns.;

## Floating Roof (Deck)

Where the tank has an **open top** and a floating roof, it is called an **External Floating Roof (EFR)**.

Where a tank has a **fixed roof** and an internal floating deck, it is an **Internal Floating Roof (IFR)**.

The purpose of the deck is to reduce evaporation losses and air pollution by reducing the surface area of liquid that is exposed to the atmosphere. The lack of headspace ensures there are no 'breathing' losses from the tank.

The deck is a disk structure with sufficient buoyancy to keep it afloat under all design conditions. There is a clearance between the deck and the tank's shell, so that it does not bind as the deck moves up and down. A flexible seal is inserted into the gap between the floating deck and the tank shell to reduce the area of liquid in contact with the air.

### *External Floating Roof*

EFR tanks will typically have one of the following styles of floating roof:

- *Floating Pan*-no built-in buoyancy, just a single deck in contact with the liquid. Low cost but vulnerable to sinking or capsize;

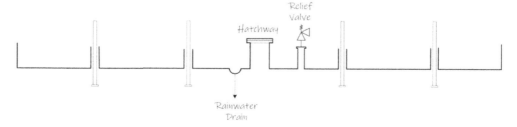

- ***Single Deck-***cheaper to construct than a double deck as only a single skin (in contact with the liquid) is required. Suitable for high vapour pressure liquids. In the event of the deck flooding, buoyancy is maintained by an annular pontoon built on the top of the deck. Tends to be weak structurally with low stiffness. Not suitable for very large tanks;

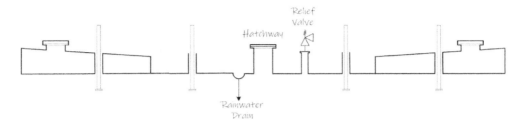

- ***Double Deck-***top and bottom decks are welded together to form a floating pontoon, with independent buoyancy chambers to provide a high level of redundancy. Suitable for high vapour pressure liquids. Structurally very strong. More expensive than a single deck to fabricate but it is more suitable for very large open top tanks, storing products such as crude oil.

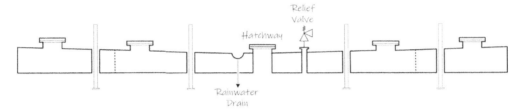

### Internal Floating Roof

- ***Floating Pan-***same as for EFR tanks. No built-in buoyancy, just a single deck in contact with the liquid. Low cost but vulnerable to sinking or capsize;

- ***Aluminum Internal, Non-Contact-***buoyancy comes from aluminum pontoons floating on the top of the liquid, with a single (non-contact) aluminum deck mounted above the liquid, creating a vapour space between the liquid and the deck. Suitable for high vapour pressure stocks. This is the cheapest internal design and can be installed retrospectively through the shell manway. Rapid field installation but weak structurally. Material of construction limits which liquids it can be used with, and it is not suitable for tanks with high turbulence. Has a shorter service life than steel;

- ***Aluminum-Internal, Full Contact-***closely linked double skin aluminum pontoons floating on top of the liquid, eliminating the vapour space found on 'non-contact' variant. Pontoons are sized to enter the tank through the manway, so can easily be retrofitted. Easier to repair than 'non-contact' and less liable to sinking or capsize. Material of construction limits which liquids it can be used with, and it is not suitable for tanks with high turbulence. Has a shorter service life than steel;

✳ ✳ ✳

**FIG. 101A** Cone Roof tanks used to store non-hazardous products. Tall tanks are frequently used where the product is heated or stirred. They are useful in small, odd shaped plots such as this.

**FIG. 101B** Dome Roof tanks. These have supporting columns holding up the roof trusses.
Photo: Tank Storage

**FIG. 101C** Aluminium geodesic roof being assembled on the top ring of a tank being constructed using the 'jack-up' method. The bottom plates of the tank are visible under the dome structure.
Photo; YHR Tanks

**FIG. 101D** Typical External Floating Roof tanks. The roofs are full width, full contact double decks.
Photo; MesaETP

## Floating Roof Components

Where the tank has an internal aluminum deck, fittings and fixtures are usually simple. External floating roof tanks, where the deck is usually made from steel, are frequently more complex and usually incorporate the following features:

- **Roof drain** – if the deck is exposed to the weather it must be designed to deal with rainfall or it could capsize or sink. Decks are usually designed to fall to a central drain sump. The drain is connected to either a flexible hose or metal pipe with articulated joints that go down through the product to the bottom of the tank. In the event of rain, the water drains to the centre of the roof and down to grade (via the flexible or articulated drain lines). When everything works as designed, the rainwater is not contaminated by the product within the tank, or the tank contents contaminated by the rainwater. The rainwater drain nozzle has a valve; in theory, it is usually assumed the drain valve is kept closed until it is required to drain water from the floating roof. In practice, however, this valve is usually left open (tut tut).

  Floating roof drains always present a risk to the integrity of the tank. If the drain valve is not opened, water could build up on the floating deck, reducing its buoyancy. If the valve is left open, any failure of the hose, or an articulated joint, could result in a significant portion of the tank contents being discharged into the contaminated water drain system;

- **A hatch to manually 'dip' the tank** – these hatches are not frequently used as tanks of any significant size have more sophisticated ways of measuring the tank contents;

- **Vacuum breaker valve** – to prevent physical damage if subjected to a partial vacuum;

- **Deck legs** – when the tank is inspected internally, it is not desirable to land the deck onto the tank bottom plates as this will hamper the inspection. It is therefore normal for the deck to have legs that can be raised or lowered. The legs are normally left in the raised position until required. If left in the lowered position they will reduce the operability range of the deck and possibly increase the tank 'heel' or 'deadstock'.

- **Anti-Rotation post** – to stop the deck from rotating within the shell, a vertical post is frequently installed;

- **Access gangway** - leading from the top of the shell to the deck, for Operator access. These gangways are fixed with hinges onto the top of the shell and designed to slide across the top of the roof as it rises or falls with changes in the liquid level. It is common for these gangways to feature self-leveling stairs.

- **Fire protection equipment**- considered in "**Chapter 4-Loss Prevention**";

- **Earthing straps** – between the deck and the shell

- **Rim seal** – to reduce vapour loss from the fluid in the annular gap between the deck and the shell. Rim seals come in a variety of forms;

  ❖ Resilient foam, sometimes covered with a rubberized fabric, either in contact with the liquid or above the liquid;

❖ Wiper blade, usually rubber or foam, above the liquid;

❖ Mechanical seal – usually where a metallic shoe is pressed against the inside of the tank shell and the top of the shoe is connected to the deck using a rubberized fabric to act as the vapour barrier. Contact between the shoe and the shell is maintained by a weighted cantilever or a spring.
In many cases wiper blades are installed as secondary seals, in addition to additional weather protection mounted off the top of the deck to prevent the deterioration of the rubberized fabric;

## Tank Bottom

The factors that dictate the design of the tank bottom are:

- **Water contamination**-Tanks 'breath' as the temperature rises and falls during the day. The air entering the tank contains water vapour, which condenses within the tank. In addition, tank roofs (and floating decks) are seldom entirely watertight, so rainwater can find its way into the tank. Where products are transported by sea or river, contamination of the cargo with seawater is not uncommon;

- **Sediment**-some commonly stored liquids, such as crude oil, are unrefined and sediment can settle out while the tank is in service. When the tank is removed from service the sediment has to be removed to allow the tank to be inspected;

- **Changes of product**-some tanks operate in a 'swing' role, meaning that they may be used for a variety of products. To avoid cross-contamination, these tanks have to be capable of being fully drained before a different product is introduced;

- **Product quality**-some products react badly to prolonged exposure to water in the tank. The principal example is aviation fuel, so frequent and thorough decanting of the water and fuel/water scum at the bottom of the tank is necessary.

The introduction of vapour recovery systems has reduced the ingress of water due to tank breathing where volatile liquids are stored. However, water inside the tanks remains the primary cause of internal corrosion, and therefore, tank failure.

The removal of water contamination from the tank is greatly aided when the tank is allowed to 'settle'. As the water and the product usually have different specific gravities, over time they will separate.

An old rule of thumb within refineries is that you need at least 3 tanks for each product. One tank will be on 'rundown' duty, accepting newly refined products from the process units. One tank will be full and 'settling' and the product is being tested to ensure compliance with the product specifications. The third tank will be used to supply 'clean' and on spec. product.

- **Flat bottoms** are mostly fitted to small tanks (typically 10m in diameter or less) and are widely used in the chemical industry. They are simple and economical to fabricate and install. However, they are difficult to drain, so not suitable for frequent changes of products where contamination must be avoided.

6

- **Flat Bottoms with a Single Slope** are similar to flat bottomed tanks but can be drained fully and can be up to about 30m in diameter. A floor drain is installed at the lowest point of the tank and a concrete catch pit is formed within the tank foundation, meaning the drain can be inspected for leaks from outside the tank. Can be suitable where large quantities of sediment are expected, as they are easier to clean out.

- **Cone Up** is the most commonly used type of tank bottom and is frequently found in the petroleum industry. Water is less likely to be found between the bottom of the tank and the foundation, leading to reduced corrosion of the bottom plates. Cone-up tank bottoms drain toward the perimeter of the tank. If a sump is installed at the perimeter, the tank can easily be drained. When used with products having an SG of more than 1, any water in the tank sits on top of the product. This means that a takeoff at the bottom of the tank takes an uncontaminated product.

- **Cone Down** is suitable for refined products where minimum contact of the product with water is acceptable. Being cone down, liquid drains towards the centre of the tank, where a drain line is welded into the bottom plates of the tank.

  Seldom used now as the drain line is prone to damage and difficult to inspect. A failure of the drain line cannot easily be detected and could lead to the loss of the tank contents.

- **Cone Down with Sump** is similar to the cone down design, but with the refinement of a central sump. Seemingly a minor change, the central sump, and revised foundations allow any leaks from the drain or the tank bottom to be easily spotted and allow rectification work to salvage the situation. The small central sump also minimises the surface area of water in contact with the product.

  Cone-down tanks are used with products having an SG less than 1, meaning that the product sits on top of any water in the tank. The drain line to the central sump can be used to draw off water from the tank. Any fine sediment in the liquid will, over time, find its way towards the floor of the tank and settle out. Excessive quantities of sediment may accumulate in the sump, but draining water will also help remove the sediment.

## 1.2 Tank Design

**Standards**

Within the European Union (EU) the specification for the design of such tanks is covered by BS EN 14015. Within the USA, API 650 is used.

Whilst these codes are often used internationally, many countries have their own design codes, frequently based on the BS (EN) and API documents, but with minor modifications and variations.

The most common tank design standards are:

| | |
|---|---|
| BS 2654 | Specification for Vertical Steel Tanks (old-withdrawn)) |
| BS EN 14015 | Specification for Vertical Steel Tanks |
| API 620 | Design & Construction of Large Welded Low Pressure Tanks |
| API 650 | Welded Tanks for Oil Storage |

| API 651 | Cathodic Protection for Above Ground Tanks |
| API 652 | Linings for Above Ground Tanks |
| API 2000 | Venting Atmospheric Tanks |
| API 2350 | Overfill protection for Storage Tanks |

## Material Selection

Tanks are large items and represent a significant capital investment on the part of the Owners. Even a small change in material selection or design code can have a significant impact on the total installed cost.

It would be convenient to think that tank construction materials are selected based on the lowest price consistent with the specified tank duty. However, there are many factors driving the choice of materials for construction:

- **Engineering issues** include;
  - ❖ Material selection;
    - o Material cost;
    - o Availability of material locally;
    - o Availability of suitable fabrication facilities;
    - o Ease of fabrication;
  - ❖ Corrosion prevention;
  - ❖ Design life and corrosion allowances;
  - ❖ External loads including seismic;
  - ❖ Cathodic protection;
  - ❖ Leak detection;
  - ❖ Blanketing system;
  - ❖ Linings and coatings;
  - ❖ Style of tank roof/bottom required;
- **Liquid to be stored** and its physical properties including;
  - ❖ Vapour pressure;
  - ❖ Flash point;
  - ❖ Operating pressure;
  - ❖ Heating/cooling requirements;
  - ❖ Specific gravity;
  - ❖ pH;
  - ❖ If food contact material certification is required;
  - ❖ Potable/de-mineralised water;
  - ❖ Limits on contamination/turbidity;

The majority of ASTs are constructed using:

- **Carbon Steel** - Also referred to as mild steel, this is the most common material for tank construction. It is readily available and can be formed, fabricated, and machined easily. A downside is that carbon steel is also prone to corrosion, which means it is necessary to coat/paint the outside of the tank. In some situations, it may also be necessary to coat the inside of the tank, but this is usually to avoid product contamination.

The design codes (like API 650) contain a list of permissible steel plate standards such as ASTM A36, A283, etc. for the designer to select. As each steel plate standard has its own physical properties, once the material is chosen the design codes are straightforward and quite specific.

- **Stainless Steel** - Where the liquid being stored is corrosive, or where the tank is for chemical storage or products for human consumption, austenitic stainless steel is used. There is a significant price premium over carbon steel but stainless steel does not require a coating to prevent corrosion, does not contaminate the product, and can withstand the high temperatures associated with steam sterilization and cleaning.

- **Reinforced Plastic** - May be used where the resistance to chemicals is acceptable. Their lack of fire resistance means they are not used for storing flammable or combustible liquids. Tanks made from reinforced plastic are usually made in a specialist fabricator's yard and delivered to the site as a complete assembly. As an alternative, the reinforced plastic sections can be bolted together at the site if a larger capacity is required.

- **Aluminum** - Historically used for cryogenic applications because aluminum remains ductile at lower temperatures than carbon steel. However, stainless steel tanks have now taken over this market.

There are a variety of alternative materials, such as glass-coated carbon steel plates (suitable for bolting together on site) for specific duties such as potable water and water treatment. Large tanks designed to hold chemically aggressive products, such as alkali or acid, may be lined internally with a resistant coating (frequently an epoxy or polyurethane material sprayed or brushed on). However, these are not generally resistant to abrasion and have to be re-lined regularly to maintain the coating integrity. Lining materials can have rust-stabilizing and corrosion-inhibiting properties if specified.

A danger with internal coatings is that if there is a fault in the lining and any area of the metal is not fully coated, the corrosion at that point is extreme. To avoid this type of pin holing, the coatings are usually 'holiday tested' during inspection to find any fault not obvious to the naked eye.

## 1.3    Tank Erection

**Preparation**

Standard practice in the industry is for a single contractor (the Tank Designer/Erector) to be responsible for design of the tank, procurement of materials, prefabrication, delivery to the site and erection of the tank, using their own labour. A Major Civil Works Contractor will frequently design and construct the tank foundations, the secondary containment, the floor of the secondary containment and any associated drainage systems.

The Tank Designer/Erector will choose the tank plate material and thickness, depending on what is available in the market at that time. There is no standard size of plate used for tank construction, but generally, they will be between 2.4m and 2.8m wide. The design cannot be completed until the material and plate sizes are known.

Plates are ordered and delivered to the Erector's fabrication yard. Each course will have a different thickness as the hydrostatic pressure at the bottom of the tank increases as the tank gets taller. Plates to be used for lower courses in the tank shell will be thicker than those at the top of the shell.

The plates are inspected for defects, usually using ultrasonic methods to look for internal discontinuities. If passed, the plates are cut to size, depending on where in the tank they are to be used. The plates for the shell will have bevels or profiles cut along each edge in preparation for welding at the site. The shell plates are then rolled to achieve the correct profile before delivery to the erection site.

## Tank Roof

The greatest risk to tank constructors is working at height and falls are the most frequent cause of death. It is much safer and quicker to build the roof at ground level as soon as the floor plates are laid and welded.

The roof of a cone roof tank built 'conventionally' is constructed on the free ground to one side of the tank foundation. The roof's supporting structure is built and covered in steel plate. When the tank shell is completed, the roof is lifted into place using a crane.

FIG. 102 The roof structure can be lifted into place as a bare structural skeleton, or be fully plated out prior to lifting.
Photo; TIW Steel Plates Inc.

The roof of a cone roof tank being built using the 'jack up' method is erected on the top two courses of shell and jacked up with the shell plates as construction progresses (see below for more information on the different methods of building tank shells).

The deck of a floating roof tank is built on the tank floor, and the tank shell is erected around the outside of the roof. Rim seals are installed after the hydrostatic test.

## Tank Floor

The floors of small vertical cylindrical tanks are generally constructed from steel plates, butted together, and seal welded. The joints are staggered and the shell sits directly onto the floor plates. This is possible as the hydrostatic pressure on the bottom plates is limited as it is a small tank.

Most large vertical cylindrical tanks have floors consisting of steel plates, overlapping and joined together using fillet welds. Some designs use a layer of bitumen and sand mixed together for the floor plates to be laid on (either on top of the concrete or the compacted sand).As the plates are welded together the heat from the welding helps the plates bed into the bitumen-sand mix, filling the voids under the plates to prevent water ingress. Around the perimeter of the tank is the annular ring, made out of a thicker steel plate. The purpose of the annular ring is to carry the additional weight of the shell and, if necessary, the additional weight of the roof.

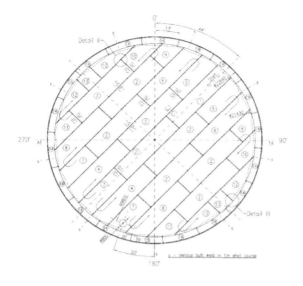

The shell/annular ring weld is the most critical weld for any tank as the failure of this weld will result in the total loss of the tank contents

Electric arc welding is used to make the fillet and butt welds including the shell to annular ring weld. Inspection of the fillet welds is usually done with a vacuum box and a soapy solution. In the event of weld porosity, air is drawn from below the tank and the soapy water creates bubbles in the glass-topped vacuum box

## Shell-Traditional Method

The 'conventional' method involves building the tank by stacking one plate on top of another.

First, the bottom course is erected, and then the plates for the second course are lifted into place above the first course. The curved shell plates have matching lugs at the corners. Steel wedges are driven into adjoining lugs that hold the plates in position with the ones on both side and the plates below. In this way, the tank is built, course on course until it attains its full height.

FIG.103 Conventional Plate-on-plate construction. Access platforms can be inside or outside the shell but will be dependent on the method of welding Photo; TIW Steel Plates Inc.

As the tank grows, scaffolding is erected around it so that the welders can butt weld the horizontal and vertical seams. Sometimes temporary staging is used instead. In some parts of the world, this staging can be very sketchy.

With cone roof tanks, when the tank shell is full height and fully welded out, the roof is lifted and attached to the top of the tank. The floating decks of open-top tanks do not need to be lifted, as they will rise as the tank is filled with water during the hydrostatic test.

*Advantages*

- *Method of construction suitable for all diameters and heights;*
- *Possible to build with greater accuracy;*

*Disadvantages*

- *Long duration of erection/construction;*
- *Working at heights;*
- *More resources are required;*

## Shell-Jacking Method

This method avoids the majority of the work 'at height'. The top two courses of the shell are erected conventionally and fully welded, including any wind girders or reinforcement on the top course of the shell.

The cone (or dome) roof is assembled on the curb ring. Supporting steelwork, roof plates, and access platforms are installed to produce a finished roof before the tank shell construction continues.

Load-bearing lugs are welded to the inside of the shell plates and jacks are installed at each lug to lift the shell. Although the number of jacks is dependent on the total weight to be lifted, you can expect to see a jack every 3m or so.

A hydraulic power pack and oil reservoir are installed at the centre of the tank foundation and hydraulic hoses run to each jack. The roof and top shell rings are hydraulically lifted to a height where the next course can be inserted below them.

This new ring is welded together and to the rings above. New lifting lugs are installed one course down from the original level, and the jacks are moved to coincide with the new lug positions.

The shell rings/roof are lifted hydraulically to a height where the next course can be inserted below them. The spiral stairway (with handrail post) is installed as work progresses and welded with its support bracket and the stair stringer.

FIG. 104   Jacking up the roof and the shell. When first seen, the method appears inherently unsafe, but it is well proven and, in competent hands, is safer than the traditional plate-on-plate erection method. Photo; Bygging-Uddermann

The process is repeated until the tank is up to its full height. The jacks are removed, the weld between the bottom course and the annular plate is completed and the tank is filled with water for a hydrostatic test.

### Advantages

- *Work at ground level- safer with lower risks, time-saving and economical;*
- *Better productivity, better quality when not working at height;*
- *Wind damage to shell while erecting is eliminated by installing roof/wind girder first;*
- *Tank plumb reading within API 650 tolerances is easily achievable;*
- *Less involvement of high-capacity cranes;*
- *Scaffolding costs are held at a minimum;*
- *The air gap between the tank bottom and the bottom of the shell reducesthe effects of the wind load;*

### Disadvantages

- *Cannot be used for the erection of column-supported cone roof tanks;*
- *Only single-sided welding can be achieved;*
- *Not be feasible for double wall tanks;*
- *For tanks less than 15meters in diameter, the jacking up method does not give any significant advantage;*
- *For tanks more than 90meters in diameter, the jacking up method has practical limitations;*

## Shell-Coil Method

A relatively new method of construction but is only justified when several small tanks are to be built at the same time. It is a variant of the 'jack-up' technique but instead of using rolled plates, the material is supplied as a large continuous coil. As the material comes off the coil, it goes through rollers to give it a curve and enough material for a complete ring is formed in one operation. The ends of the ring are welded together, and the rings are stacked until the required height is achieved.

This method is only suitable for small-diameter tanks and is limited to 10 or so rings. Invariably the material is stainless steel and the controlling factor is the thickness of the metal to withstand the hydraulic pressure within the tank. The use of this technique enables small tanks to be built at site quickly as there are no plates to procure and roll. However, the initial setup costs to produce 'coil' tanks are high so the number of tank builders offering this technique is limited.

## Hydrostatic Testing

Following the completion of a new storage tank, one of the final mechanical tests recommended is a hydrostatic test. The first and most obvious reason to fill the tank with water is to ensure the tank has no leaks. The tank is filled with water at a prescribed rate and then examined after 24 hours to ensure there are no cracks, pinholes, or other discontinuities in the welds.

The second purpose of the test is to confirm the mechanical strength of the tank. If there are weak spots they will be evident when the tank is full of water, with far less danger to workers and the environment than if the tank is full of a potentially toxic substance.

The final, and most critical observation made during a hydrostatic test is the integrity of the foundation that supports it. While the tank is full of water settlement can be observed and measured to provide baseline measurements for comparison over time.

Depending on the size of the storage tank a very large quantity of clean water may be required for the hydrostatic test. Where several tanks are being built the usual approach is to sequence the construction so that hydrotest water is passed from tank to tank, reducing the total volume of water required to the bare minimum.

Construction sites are sometimes remote and trucking in water is not feasible. The disposal of water following the test is also an issue as the test water is dosed with biocides and corrosion inhibitors.

API-650 does have an option to forego the hydrostatic test if additional testing is performed. A leak test must be performed using a penetrating oil test or an air pressure test. The tank must also receive additional radiographic examination to capture all weld intersections in the shell. This increases the amount of radiography by about 4 times.

In addition, API-650 requires the foundation to be over-designed. Settlement of the foundation can still be observed during the first fill with product and baseline measurements taken.

## 1.4   Tank Repairs

Most tanks are repaired after they have been taken out of service for inspection and must be:

- Gas freed;
- Any bottoms, sludge, water, or residue removed and disposed of;
- Cleaned;
- Scaffold erected internally for access;

### Tank Roofs

#### *Cone or Dome Roofs*

Tank roof corrosion can be summarised as;

- External corrosion to roof plates due to rainwater pooling;
- Internal corrosion to roof plates and supporting steelwork due to condensation;

Repair work is straightforward. Supporting steelwork can be cut out and replaced with new where necessary. Alternatively, additional steelwork can be installed to supplement the existing (diminished) steelwork. This can be done by workmen working internally or externally, depending on whether the roof plates are to be replaced, or if access internally is difficult.

Roof plates can be cut out and replaced with new plates. In some cases, it may be decided to 'over plate' the area but this can only be done where rainwater pooling can be avoided in the future. Over-plating is the cheaper option and may be the preferred choice where the tank has a limited remaining service life. It can also be done with the tank in service as long as the tank doesn't contain volatile liquids.

## Floating Roofs

Where a floating deck is installed in a tank with a fixed roof, few repairs are required. The top of the floating deck is not exposed to the weather and the bottom of the deck is usually in full contact with the stored liquid, excluding air and therefore minimising corrosion.

A floating deck in an open top tank is prone to damage from the environment. In the worst case, there can be metal loss on the top of the deck and inside the pontoons. Whilst the pontoons are usually sealed, rainwater can usually find a way in; especially if there have been problems with the deck rainwater drains. However, this usually leads to surface corrosion rather than any major structural problems. If the plating is seriously corroded it can be replaced or over-plated with fresh steel sheets. The decks of open top tanks take a severe beating from the sun in warmer locations, and repainting the deck is a typical preventative action during statutory outages.

The most common faults on open top tanks are associated with rainwater drains. When rainwater falls onto the floating deck, it is supposed to drain to a central position. However, it is difficult to fabricate large welded structures with shallow falls and rainwater pooling can occur. In environments where the rainfall is frequent and temperatures are high enough for the pooled water to evaporate, these constant wetting events can cause quite severe corrosion.

At the central location (or, on large decks, several low points) the rainwater is collected into a gulley or sump, and discharged by either a flexible hose or an articulated pipe.

Faulty rainwater drains tend to allow rainwater into the stored product. However, in some circumstances, the product may find its way into the rainwater drain. It is good practice for the rainwater drain valve to be closed at all times, other than after rainfall. The Operator should always check that only rainwater is coming out of the rainwater drain line!

Repairs to the drains usually involve the replacement of a damaged hose or new seals on the swivel joints of the articulated pipe during the scheduled out-of-service inspection outage.

Somehow, although I have never understood why it happens, it is not uncommon to have to repair the landing legs on the underside of the deck. Invariably, the Operators manage to bend a leg every time they 'land' the deck for inspection. Also, a frequent repair is where the flexible gas seal (the rubberised fabric between the top of the sliding shoes and the floating deck) requires replacement.

## Tank Shells

External corrosion of the tank shell is reasonably rare, as, even with a degraded paint system, rain does not collect anywhere. The one exception is the joint between the shell and the annular plate. When constructing the tank, the heat from the welding process tends to bend the annular plate on the outside of the tank upwards. In service, rainwater can collect between the annular plate and the shell and cause corrosion, especially on heated tanks. There are propriety products (like Belzonna) that are produced specifically to remedy this situation with either new or old tanks.

Internal corrosion can fall into two categories.

It can be light and general or local and heavy. 'Light and general' is not normally a problem as the tank shell has a corrosion allowance built in and the wall thickness of the shell is assumed to reduce progressively. The worst situation for a tank shell is with pitting corrosion, where there can be a significant reduction of metal thickness in a small localised area.

This type of corrosion is frequently seen in tanks storing crude oil, where a band of corrosion around the tank about 1 or 1.5 meters above the floor plates can often be observed.

In severe cases, the only solution is the replacement of the damaged sections of the steel plate by flame cutting out the damaged section and welding in a fresh steel coupon of the same dimensions. In extreme cases, the whole steel panel will need replacement.

## Tank Floors

Tank floors are frequently made up of lapped steel plates and the joints are secured and sealed using fillet welds. In service, the fillet welds can fail, leading to the bottom plates leaking. In this case, grinding out the old weld and re-welding the plates usually solves the problem.

Corrosion of the bottom plates can be internal or external. In most cases, the bottom of the floor plate is sitting directly on the foundation, which can be a variety of materials. Although most tanks will have sacrificial anodes or suppressed current protection, water trapped between the foundation and the underside of the tank will cause accelerated corrosion.

The corrosion can also be caused by the liquids inside the tank. Heated tanks, using steam heating coils, are particularly prone to corrosion if the coil leaks steam/hot condensate into the bottom of the tank. Also, with products with a specific gravity of less than 1, any condensation from the atmosphere inside the tank will eventually find its way into contact with the bottom plates. Although flat bottom tanks are more prone to this damage, it can also happen with 'cone up' and 'cone down' tanks.

The bottom plates are thin compared with the shell plates. In the event of excessive corrosion of the bottom plates from the inside or the outside, it may be necessary to replace some, most, or all of the bottom plates.

The bottom plates are easy to replace, but to get them into the tank it is necessary to cut a shell plate out of the bottom ring. Making an earth ramp to the top of the foundation in front of the opening makes it easy to get a Bobcat (generically known as a skidsteer, wheel loader, or payloader) inside the tank to move and position the steel plates. When the floor is completed, the section of the shell is replaced and welded into place.

An alternative is to fit a 'dummy' floor inside the tank. Using the same approach as listed above, the tank is opened up by cutting an access hole in the shell. Sand is placed over the damaged floor and leveled to a depth of 150-200mm. The new floor plates are laid on the sand and welded in position. With this method, the shell doesn't sit on the new floor plates, but on the old annular ring. The new floor plates have to be trimmed to sit inside the shell and fillet welded in place. This approach has been used successfully with heated tanks, as the dummy floor effectively creates a double bottom to the tank and provides additional insulation between the tank contents and the foundation.

Another approach when the foundations need repairs is to jack the tank up with hydraulic jacks. When the tank is at a working height, wooden blocks are placed around the perimeter of the tank, ensuring it is safe for workmen to venture underneath it.

## Hydrostatic Testing

Following a "major alteration" (as defined in API-653) a hydrostatic test is recommended. However, these tests are rarely carried out when the tank has previously been in service. In most cases, the tank construction specialist carrying out the repairs can provide sufficient testing and supporting technical justification to verify the integrity of the vessel.

## 1.5 Tank Demolition

Substances not normally regarded as presenting an explosion hazard can give off inflammable vapours when heated and form an explosive or combustible mixture when mixed with air.

In some cases, sludge at the bottom of a tank may be more hazardous than the original contents. Such deposits may not always be completely removed by the cleaning process.

The hazard varies according to the type of tank involved;

- Vertical tanks - in the event of an explosion the roof will generally fail at the shell/roof weld and vent the tank;
- Horizontal or spherical tanks of uniform construction and with no lightweight section. These present a greater all-round hazard from the lateral blast;

### Venting & Cleansing

In the case of small vessels, both vapours and residues can be removed by steaming out.

With larger tanks, the problem is somewhat different. Owing to the high heat capacity of the tank, steaming out cannot be relied upon unless very large quantities of steam are readily available. It is generally an easy matter to eliminate any explosive concentrations of vapour within the tank by either forced ventilation or natural ventilation.

Volatile materials will clear rapidly, but less volatile materials may pose a hazard even after they appear to have cleared, and the problem of residual fires may remain.

Ventilation by itself will ultimately only remove the volatile materials present in a tank and will never remove heavy ends or solid residues and tars. These types of residues themselves can contain considerable quantities of volatiles and unless they are removed, give rise to a very high fire risk. Sparks and slag from the cutting torch can serve as very effective sources of ignition and fires can readily occur.

### Removal of Residues

The removal of inflammable residues from these tanks can be very costly and time-consuming.

Where the tank is for demolition and it is only of scrap value, the high cost of tank cleaning may cause less scrupulous Owners to try and avoid this necessary stage. Any competent tank demolition company will refuse to work on these tanks as their personnel and reputation are at risk.

In most cases, residues will be found at the bottom of storage tanks but in certain cases 'hang-up' can occur on side walls. In some cases, a film of inflammable material can occur over the tank walls and this material can be particularly hazardous.

## Inerting

Tanks may be filled with nitrogen, carbon dioxide, or other inert gas. In these cases, the atmosphere in the tank must be thoroughly tested to ensure that the oxygen level is below 10 percent.

A method favoured for floating roof tanks is to fill the tank with water and demolish the top ring of the tank.As the water level is lowered and the second ring is demolished. In this way, the tank is demolished ring by ring until the roof lands on the floor. Of course, all the water in the tank has to be treated before disposal.

An alternative method for fixed roof tanks is to fill themwith expanded foam. As long as the foam is fully expanded before going into the tank, and the tank is filled from the bottom, tanks can be made gas-free before demolition. To maintain the level of foam in the tank,the additional foam has to be continuously pumped in.

## Demolition

Demolition sites are inherently dangerous and are frequently fenced off from an operational area. If there is a risk of airborne hazards, such as asbestos, it is a wise precaution to have air sampling devices around the perimeter of the fenced area, or at least downwind.

Operators must make the tank safe before it is handed over to the demolition contractor. Operators will generally:

- Clean the tank and leave it gas free and open to the atmosphere. A gas-free certificate must be provided to the demolition contractors;
- Isolate all incoming and outgoing lines with a marked spade or blank flange;
- Isolate all other lines in the vicinity;
- Identify any underground services in the vicinity of the tank;

Once the tank has been made gas free and cleaned, demolition is usually straightforward with the same types of attendant risks seen in any industrial or commercial demolition.

In some circumstances, tanks have been demolished using scaffolding or mobile elevated platforms to burn out sections of the tank. However, this increases the risk to the demolition team due to working at height and manual handling issues. To minimise health and safety risks, the demolition contractor will minimise the use of labour and hand demolition techniques.

FIG. 105        Hydraulic shears make short work of bringing a tank down. Once on the ground, conventional demolition methods are used, the shell, roof and floor plates are cut to transportable sizes and disposed off site.
Photo; Technical Demolition Services Ltd.

18

It should not need saying but it is NOT a good idea to be inside the tank while it is being demolished. As obvious as it seems, it is not unknown for workmen to cut away the roof supports while standing under the roof. This is very difficult for the Engineer in charge to explain away at the subsequent court of inquiry and public prosecutions.

There was a well-known case in the UK where a crude tank was being cleaned. The tank had been pronounced gas-free although there were still a lot of residues (sludge and bottoms). To clean out the residue a team of workmen was employed to shovel out the muck by hand. Believing it was safe, the work gang would stop regularly to have a smoke break, still inside the tank. Their luck didn't hold out and a number were killed in the subsequent fire. It was never really explained how they had got a cigarette lighter inside the tank.

If a tank or pipeline is contaminated with flammable liquid, it may be necessary to cut it using pneumatic reciprocating saws to minimise the fire risk.

Tanks are very easy to 'drop' as long as you know where to cut into them. The safest method is to use a tracked excavator fitted with large hydraulic shears. With very large open top tanks, the excavator can be inside the tank (after the floating roof has been cut up and disposed of). With smaller tanks, the excavator sits outside of the tank.

Shears will cut through the metal skin of the tanks in a specifically designed sequence and allow the walls and roof of the tank to be folded in, whereby the sections can then be cut into optimum sizes for recycling. After the roof and shell have been cut up and disposed of, the tank floor plates can be cut up. Consideration must be given to the possibility of any remaining product underneath the tank floor plates, which has collected over years through any leaks that may have been present.

Through this process, there is a risk of flammable vapours or pockets of flammable liquids. It is quite common on many demolition sites to see a welder with a gas torch in one hand and a fire extinguisher in the other.

## 1.6 Other Types of Tank

**Small Capacity**

### Rectangular Bolted Modular Tanks

Assembled using standardized square panels and bolted together. The square panels are frequently galvanized steel, coated steel or plastic and the tanks are generally used for water storage. They are ideal where the liquid being stored does not pose any environmental risks, and where only semi-skilled labour is available to carry out the assembly work. They are usually mounted on a steel sub-frame:

### Single Wall Tanks-

Small capacity single wall tanks are either vertical or horizontal. Horizontal tanks are usually supported by 2 saddle supports. Where the contained liquids pose an environmental or health risk, the tanks are enclosed within a steel or concrete bund as secondary containment;

### Double Wall Tanks

These have become common for above and underground applications since the outer tank contains

19

any leak from the inner tank. In some instances the space between the inner and outer walls of the tank is filled with an inert fluid. In the event of theinner wall leaking, the product will displace the fluid, triggering an alarm.

These types of tanks are generally made using GRP or fabricated with various plastics including HDPE. Above ground tanks can be vertical or horizontal, but underground tanks are only horizontal. It is common for horizontal tanks to be weighed down to counteract any buoyancy from the water table.

## Bolted Tanks

Frequently fabricated from glass-coated steel or plastic sections, the tanks are bolted together on site. They are ideal for bulk water storage and due to the chemical resistance of the materials used may be suitable for potable water, seawater, fire water, or water treatment facilities.

They are usually flat bottom tanks, sitting on a simple flat concrete base, and may be open or closed roofs. Due to their low initial cost and low maintenance requirements, these tanks are used in agricultural areas for water storage and anaerobic digester systems.

## Cryogenic tanks

Suitable for the storage of liquefied hydrocarbon gases (LHG) and liquefied natural gas (LNG), liquefied commercial gases such as oxygen and nitrogen, and refrigerated liquids. These products are cooled until they are liquefied and stored at (nominally) atmospheric pressure. Typical storage temperatures are -196$^0$C for air and -160$^0$C for LNG.

In practice, the liquids tend to boil off and internal pressures can increase. As a result, the products are sometimes kept in pressure vessels or additional cooling systems are required to keep the product temperature low. In every case the tanks or vessels require insulation.

For flammable products, fire codes stipulate secondary containment which is usually concrete. At low temperatures, metals can be prone to brittle failure so material selection and strict adherence to design codes are required.

Cryogenic tanks are a class to themselves and present problems and constraints not typical of a normal terminal. Therefore they will not be included in the following comparisons or explanations.

## Oddities

The thing about Tanks and Terminals is that they always have the capacity to surprise.

There is the crude oil storage that was constructed underground in the middle of a national park near Marseilles. The caverns were solution mined into a salt layer deep underground. This was done by drilling two wells 800-1000m deep. Water was pumped in through one well and out through the other. Over time the salt was softened and washed away, forming a cavern. By adjusting the depth of the inlet and outlet pipes, caverns can be created up to 200m across and 400-800m deep depending on the geology. This particular Terminal consists of several dozen caverns and they are used for the strategic storage of oil. As the crude is ultimately going to a local refinery, the contamination by the salt is not an issue (it can be removed during the refining process).

A similar process is used to construct the caverns for the storage of gas underground. In the case of natural gas, it is under pressure so the caverns can only be sited where there will be a protective rock cap over the cavern. The gas only needs to be dehydrated when it leaves the caverns, which cannot be fully drained after construction so have some water in them.

Then there are the storage tanks created in a time of war.

During World War II, the UK built a series of Terminals across the country to store aviation fuel. These Terminals were widely dispersed across the countryside and hidden from enemy eyes to avoid being targets for German bombers. The tanks look like standard, atmospheric, vertical, and cylindrical tanks but are buried. Instead of a steel roof, the tanks have 1m thick concrete roofs that would withstand anything other than a direct hit and the whole tank farm was covered with topsoil. Eighty years later those tanks are still there and many are still in operation (partly because to demolish them would be anHSE nightmare).

Storage tanks protected from air attack can be found in many places, usually as concrete above-ground tanks inside a concrete containment vessel. In France, they are near the old submarine berths and in Cairo they are at the old refinery. Clues to the purpose of the structures are the age and the fact that they are made entirely out of concrete far thicker than necessary.

## 1.7    Foundations

### Types of Foundation

There is a wide variety of tank foundation designs. The more common types are:

* Raised pad of compacted fill (sand pad):
* Concrete/crushed stone ring wall with compacted fill:
* Engineered fill (ground improvement):
* Concrete slab (raft):
* Piled foundation:

As with most things related to tanks, foundations can be a composite of various design elements, as necessary to suit the site ground conditions and tank design.

There is no simple or straightforward way to decide which style of foundation is the best for a specific location. In some cases, the design will be dictated by the available materials, local practices, local regulations, physical constraints or contractor experience.

At a new site, the bearing capacity of the soil can be improved by draining the soil, compacting the soil, replacing the poor soil or stabilizing the soil with grout or chemicals.

### Foundation Defects

After commissioning, problems with tank foundations are rare, but they generally fall into the following categories:

* Settlement
* Differential settlement
* Ovality
* Buckling

Tank construction contractors are adept at blaming all tank defects on the design/construction of the foundation (which has invariably been done by a third party). As a rule of thumb, the more experienced the tank contractor is, the less likely that any defect will be apparent during commissioning.

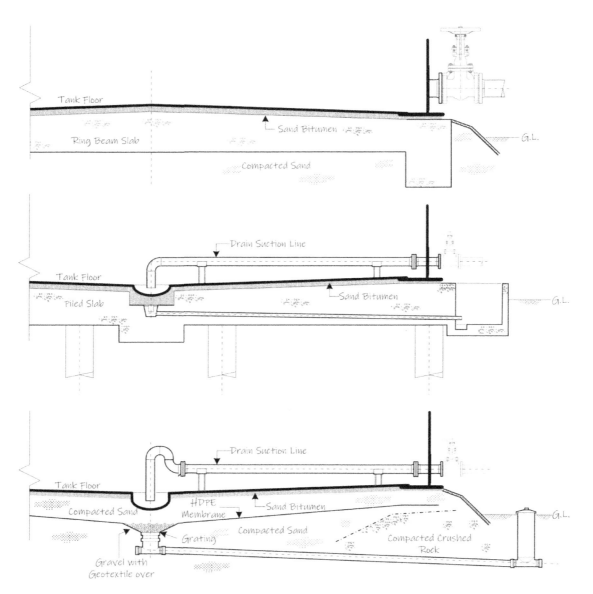

DIAG. 101    Three different types of foundation design. A cone down tank can just as easily be built using a concrete ring beam and compacted sand infill. A cone up tank can be built on a compacted crushed rock ring. The drain line/gulley at the bottom of the foundation is to give an indication of leakage from the tank bottom. In this example, leakage on a cone up tank will be evident at the edge of a tank.

22

Tank contractors have their techniques for solving problems. The Owner's Engineer supervising construction should generally be absent from the site while the corrections are made and not enquire too closely about how it was achieved!

There is sometimes a strong case to be made for plausible deniability.

The worst tank foundation problems usually happen suddenly and without warning (example-leaking bottom plates over a long period can make the foundation unstable, but the first indication may be shell buckling and a total collapse of the tank).

## 1.8    Bunds & Dykes

The tank is the primary containment. Legislation in most countries stipulates there be **secondary containment** for many stored liquids. The only liquids not requiring secondary containment are those that don't pose any environmental, health, or safety risk.

We have used the following definitions:

- **Bunds** are usually concrete-lined pits or free-standing walls:

- **Dykes** (also known as a dike, embankment, berm, or levee in American English)-are liquid retaining structures consisting of earth, sand, or clay mound. Frequently has a concrete cap to eliminate erosion, weathering and slump, and to protect the dyke in the event of a fire;

Common terminology is that tanks are installed in a **tank pit** (even when the tanks are above ground). Tank pits have bunds or dykes to retain the liquid in the event of leakage or tank collapse.

### Design Standards

Bunds and dykes should be liquid retaining structures although few (if any) are built to established national and international standards. Design standards have been made stricter in the EU due to the fire at Buncefield and the subsequent investigations.

- BS EN 1992-3          Design of Concrete Structures
- BS8007               Design of Concrete Structures (withdrawn)
- CIRIA 163            Construction of Bunds for Oil Storage Tanks (superceeded)
- CIRIA 736            Containment Systems for the Prevention of Pollution

If your work includes secondary containment, you should read the report of the **Buncefield Major Incident Investigation Board**, which explores the timeline and causes of the explosion and subsequent fire. Both CIRIA documents listed above are also worth reading as they go through the background and logic of current best practices for secondary containment.

### Retention Capacity

Industry standards are the **110%** rule and the **25%** rule.

The 110% rule is applicable where there is only 1 tank inside the secondary containment i.e. the bund needs to have a capacity of at least 110% of the primary containment volume. For example, if a tank has a capacity of 100,000m³ the bund needs to have a capacity of 110,000m³.

For bunds that house multiple tanks, the calculation is slightly more complicated. The bund needs to have a capacity to hold either 110% of the largest tank or 25% of the total volume of the tanks in the pit, whichever is greater.

In addition, secondary containment to meet current standards in western nations requires increased freeboard at the bund wall to allow for overtopping following a tank collapse. There may also be an allowance added to the capacity to contain firefighting foam and fire cooling water sprays. Typically, the bund capacity is increased by 10% to allow for these requirements.

The requirements are usually written into national design codes and the Operating Permit is usually awarded only to a new Terminal meeting the standard applicable at the time.

Taller walls will increase the capacity and protect against the tidal wave effect. However, constructing taller walls can cause access issues, limit ventilation and make firefighting difficult. For this reason, walls should be constructed no taller than 2 meters wherever possible. To minimise the tidal wave effect you can install deflector plates that limit the amount of liquid that could overtop the bund in the event of a catastrophic failure.

New (more onerous) requirements are not generally applied to existing facilities. Older bund structures were seldom required to meet such clearly defined requirements. In fact, it can be almost impossible to find out the design capacity of older tank bunds and tank pits.

## Bunds

A bund is a secondary containment system that is designed to capture any leaks or spillages from a primary containment such as a storage tank.

A bund consists of an impermeable floor and a set of impermeable walls to create a watertight area around a storage tank. Any liquid which escapes the primary containment is retained in the bund and prevented from escaping to the environment.

It is usually mandatory to have secondary containment on all hazardous liquid storage. Their uncontrolled release may cause environmental damage and create a hazardous working environment. If you have a pollution incident you can be prosecuted, face an unlimited fine, and will be liable for all clean-up costs under the "polluter pays" principle.

FIG. 106    Concrete bund wall, more expensive than a dyke but preferred when space is at a premium. These walls are typically in the shape of an inverted Tee, with the Tee buried up to 1 metre down.

There are few legal requirements for bunds. The "**Control of Pollution (Oil Storage) Regulations**" make it a legal requirement in the UK to have adequate secondary containment on any external oil storage over 200 litres.

## Bund Construction

The design requirements for bund walls have increased over the years as environmental standards have become more onerous.

Very old tank pits may have used a simple brick wall of modest height. The bricks would typically be 'engineering' bricks with a low water absorption making them suitable for below and above ground level. Over time the height of the wall increased as the minimum bund capacity was increased as a result of legislation. If the tank contents were chemically aggressive for the bricks and mortar, the bund may have been lined. Originally, chemical resisting paints were used but these had short practical lives. Later methods included lining the pit with epoxy resin. However, the brickwork deteriorated quicker than the epoxy, so linings tended to separate from the wall. There is still plenty of evidence of old-fashioned bunds on display anywhere old tanks still exist.

As bund walls became taller, brickwork became less practical. Concrete is widely available and impervious to most acids and alkalines. It also remains effective when subjected to high temperatures in the event of a fire.

The most cost-effective reinforced concrete bund wall design is an inverted tee shape. The bottom part of the tee is below ground level and the design is highly resistant to the overturning forces when the bund is filled with liquid.

Originally, these would be cast 'en-situ' in short sections. Later, slip forming was used. More recently, it has become popular to use prefabricated sections. These standardised sections, typically 6meters long, form part of a proprietary bund system. They can also be designed to suit the bund flooring material or installation method.

New standards require that bunds be capable of withstanding a catastrophic tank failure. This can be up to 6 times the hydrostatic pressure. Hydrostatic testing of a new bund should take place to ensure it is watertight before it is put into service.

Wherever possible pipes and cables should be run over bund walls and not through the bund walls or floor. If penetrations are unavoidable it is essential to ensure they have a watertight seal installed around the pipe/cable, the sealant material must be resistant to the product stored in the bund. On certain sites, seals will need to be designed in line with specific regulations or guidance.

As the aim of a precast concrete bund wall is to ensure the containment of liquids, the main consideration in the design phase is the sealing of the joint between the precast units. If there is a risk of fire in the bund, then additional precautions may be taken. Hydrocarbon-resistant, fireproof caulking is used to seal the joint between concrete sections, and a steel cover plate isfitted to protect the caulking from the extreme radiant heat.

## Bund Integrity

The explosion at the Buncefield (Hertfordshire) Oil Terminal caused several national standards to be revised. The fires in the bunds following the explosion revealed a wide range of construction defects. Oil, foam, and fire water escaped from the bunds and led to extensive contamination of the ground. Of particular concern was the use of large quantities of AFFF in an attempt to extinguish the oil fires. It was not recognized at the time that AFFF is particularly harmful to the environment. As a consequence of this incident, the design and construction requirements of concrete bunds (and the use of tertiary containment) were improved.

One of the issues highlighted in the Inquiry was the failure of bund wall corner joints at Buncefield. The use of spacers between the formwork (shuttering) panels also caused problems as spacers are inevitably hollow and the voids created had not been sealed properly. The importance of oil resisting sealant was also identified and if a sealant was exposed to high temperatures, cover plates should be installed.

### Bund Floors & Drainage

The floor of the containment area must retain liquid in the event of failure of the primary containment. There are many ways this can be achieved:

- GRP/Fibreglass
- Compacted clay
- Geo-membrane
- Geo-synthetic clay liner
- Concrete paving slabs
- Asphalt
- Polymer membrane
- Sand/resin mix
- Volcanic rock (for high setting point liquids such as Bitumen)

Being a structure to retain the product in the event of tank leakage or failure, they will also retain rainwater. Fixed rainwater drains run via a valve pit to a gravity drain or are pumped into an external catch pit and then flow by gravity to a treatment area before discharge. Portable pumps can also be used similarly.

### Typical Bund Defects

For concrete structures, any review of the bund's integrity should consider the following items:

| | |
|---|---|
| Corner joint construction; | Porosity, |
| Corrosion of embedded steel; | Heat resisting caulking; |
| Missing cover strip at joints; | Shrinkage; |
| Cracking/crazing; | Leakage through formwork spacers; |
| Earthquake Damage; | Spalling; |
| Fire/Heat Damage; | Subsidence; |
| Honeycombing; | Surface Air Voids; |
| Impact Damage; | Thermal Cracking; |

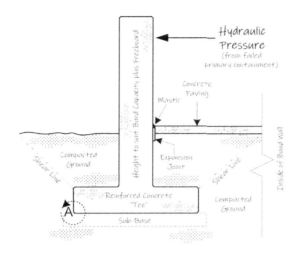

If you get the chance to wander around a construction site while the bund wall is being installed, it is worth looking for 2 items.

Frequently the 'tee' is too shallow. The logic behind the wall is that in the event of the primary containment failing, the hydraulic pressure is trying to overturn the wall (about point 'A' in the diagram). The deeper the foot of the 'tee' the more resistance there is to overturning. Too often the civil designers of the wall forget what the ultimate objective is.

The other point is the height of the bund wall. On many sites, any excavation greater than 2 meters is regarded as a 'confined space' and a PTW is required to enter it. Yet frequently Operators will go into bunds with high walls without taking adequate precautions. For consistency, many oil companies and refineries where toxic gases may be found insist the bund wall is no taller than 2m (meaning Operators can enter the bund safely and without having a PTW).

## Dykes

These were favoured for their low cost when environmental standards were not so high. Also popular for terminals and tank farms that had to be erected quickly and with small capacity tanks.

The dykes themselves tended to be quite low compared to concrete bund walls, so did not use land efficiently. Economies were also made by having a larger number of tanks in a single 'bunded' area-as only 1 tank failure at a time was considered feasible, and a large secondary containment area with a low wall was considered sufficient. With tanks spread out, ignition of adjacent tanks by radiated heat was considered unlikely.

If a fire spread to adjacent small tanks, the product inventories were small, so the impact of a fire (even a major one) was considered a risk worth taking in the drive to minimise capital expenditure.

FIG. 107    Dyke walls should have an impervious core material such as clay. This example has a concrete capping, but not all dykes are created equal!

FIG. 108    What a dyke should NOT look like! The dyke wall had slumped, and it was not able to hold the design capacity.

From the above, it may be assumed that dykes were only constructed in third-world countries. This would be incorrect as historically most secondary containment was constructed using dykes. Secondary containment built using dykes can still be seen at major terminals and refineries in Europe and America, where old tanks are installed, as the cost of installing concrete bund walls cannot be justified and there is no legislative pressure to force compliance with current standards.

## Dyke Construction

The logic behind the construction of dykes is inescapable when you consider where most product storage tanks are located.

Most refineries and major terminals rely on ships and barges to transport products or feedstock in bulk. They are therefore frequently located in estuaries and rivers with direct access to the sea. Frequently this will be on a floodplain where sediments have been deposited during flood conditions. Over time this forms a layer of alluvial clay across the estuary. Alluvial clay has low permeability and is an ideal material for the construction of dykes as secondary containment.

Because of this, the terminals and tank farms are constructed on flat open land with limited alternative uses and a ready supply of building material.

The strata of clay beneath the tanks are left undisturbed as this forms the bottom of the dyke containment and prevents the spilled product from draining away into the underlying geology. Fresh clay is excavated from elsewhere on the site (frequently during dredging for ship and barge berths) and is used to construct the dyke wall by building embankments along the edge of the tank pit. As the weight of the clay compacts the wall, it also forms a seal between the dyke and the clay forming the bottom of the 'bunded' area.

By this method, the dyke becomes a liquid retaining structure.

Frequently, the dyke walls and the bottom of the tank pit will be covered in topsoil and grass to minimise soil erosion.

A 'typical' dyke wall will not exceed 1.5m high and have a maximum incline of less than 45⁰. The dyke wall relies on the clay remaining wet to maintain its integrity and ability to retain liquids, so there has been a tendency in recent years to retrofit concrete 'capping' on the dyke to provide a degree of protection from radiated heat in the event of a fire.

The use of clay to provide an impermeable barrier to prevent the migration of contaminants is well established. It has been widely used when building housing on land with historical industrial contamination. It seems strange therefore that it was only comparatively recently that the authorities realised a fatal flaw in their logic. Before a tank is constructed it is standard practice to drill a borehole to establish the underlying geology and maximum ground bearing pressure. This information is essential for the design of the tank foundations. However, once the borehole is drilled its location was not recorded accurately and there is no guarantee that the hole was resealed (plugged) with clay. As a result, the continuity of the clay under the tanks cannot always be taken for granted.

## Typical Dyke Defects

The storage capacity of a dyke can deteriorate over time, due to weathering, slump, and erosion reducing its height. The only way to confirm the bund capacity is to have a topographic survey of the bund area. This is seldom carried out unless a specific request has been made by the Authorities to the Owner.

Some locations, especially those distant from the sea, do not have a ready supply of clay on the site with which to build the dykes. Less scrupulous builders have therefore used whatever material can be procured locally. Not only are these dykes constructed out of less-than-ideal material, but they also tend to be poorly constructed. A dyke wall that isn't compacted will be prone to settlement.

Even when new it is unlikely these dykes were ever tested to demonstrate their liquid retaining capacity. Since their construction, there may have been modifications, such as new pipework penetrations, that further jeopardise the dyke's capacity to retain liquids.

For dykes, any review of its integrity should consider:

| | |
|---|---|
| Slump | Settlement |
| Erosion | Slips/Subsidence |
| Weathering | Burrowing animals |
| Vegetation | Penetrations |
| Permeability | Physical damage |
| Scour | Vehicle strikes |

## 1.9    Other Types of Secondary Containment

### Double Wall Tanks

There is an inner and an outer tank constructed on a single foundation. The inner tank provides the primary containment, whilst the outer tank provides the secondary containment. The outer tank is constructed far enough away from the inner (but on the same centreline) to allow working space for scaffolding and the workforce during construction.

### Sheet Pile Retaining Wall

Sheet piling has many uses. It is regarded as cheap and quick, can be installed close to the boundary, and involves no excavation for wall foundations. Not generally used for the construction of bund walls (see reasons below), sheet piling is frequently used in Terminals.

On a site that slopes significantly from one end of the pit to the other, sheet piling can be erected to form a retaining wall at the high end of the pit. This can be a temporary fix during construction to hold back earth whilst the 'final' concrete bund wall is installed.

On terminals and refineries accessible by sea or river, sheet piling can be used to serve two functions. On the waterside, a row of sheet piling can form a quay wall. On the land side i.e. tank side, a row of sheet piling can form the end of an enclosed area to provide secondary containment for a storage tank. For stability, the space between the two rows of sheet piling is infilled, usually with clay dredged from the river with gravel dressing (to prevent the clay from drying out).

Over the years there have been attempts to use sheet piling in place of concrete bund walls for secondary containment. At face value, this design option could offer some practical benefits, especially with cost and schedule. However, regulators have (to date) not been convinced sheet piling should be more generally accepted as a concrete alternative.

Over the years there have been attempts to use sheet piling in place of concrete bund walls for secondary containment. At face value, this design option could offer some practical benefits, especially with cost and schedule. However, regulators have (to date) not been convinced sheet piling should be more generally accepted as a concrete alternative.

FIG. 109    Sheet piling used in the construction of a tank pit. The sheet piling is seal welded and in this case has been sprayed with concrete as passive fire protection.

This is partly due to the sealing of sheet piling. Sheet piles are typically connected using tongue and groove joints. After installation (usually by percussion piling) it is not guaranteed that adjacent piles are locked into each other and provide a liquid tight seal. To overcome this, all sheet piles would have to be seal welded along the joint after installation.

The performance of sheet piling under fire conditions is also questionable. Most research into the subject relates to the use of sheet piling in constructing underground garages. From these studies, it is clear the performance is dependent on the type of infill behind the sheet piling and the level of the water table. A bund, constructed from sheet piling and consisting of a freestanding barrier some 2 or 3 meters above the surrounding ground level, does not have this natural resistance. To overcome this, sheet piling can be fireproofed using sprayed concrete.

In a fully developed oil fire in a tank or bund, temperatures over 800 $^0$C can occur. At this temperature, the steel used for sheet piling will weaken. Even if a reinforcing beam was added to the top of the sheet piling bund, it is unclear if the piling would withstand the hydraulic shock of the product spilling out of a collapsing tank without additional stiffening or bracing.

## 1.10  Tertiary Containment

The main purpose of a tertiary containment system is to prevent the release of product and fire water to the environment in the event of a failure of both the primary and secondary containment systems. Thus, it is the third line of protection. It also allows time for additional measures to be deployed if an incident escalates.

It is not unknown for secondary containment bunds to become full in the event of a major bund fire, due to the quantities of fire water discharged to cool the tank, and foam injected to smother the fire. The preferred option, in this case, is to have a designated tertiary containment area identified so that contaminated water in the bund can be removed.

A specific risk assessment is required as there are no industry standard regulations for tertiary containment.

Options can include pumping it out to the tertiary containment area or, if the topography of the site allows, for it to flow under gravity. Ideally, the receiving area would be a separate dedicated catchment pond or lagoon.

As most sites don't have room for these facilities, an alternative is to pump the contaminated water into an adjacent tank pit. As most tank foundations extend above normal ground level, significant quantities of the discharged firewater can be accommodated within the secondary containment without it even reaching the top of the tank foundation.

Details of best current practices are contained in **"Ciria 736: Containment Systems for the prevention of pollution: Secondary, tertiary and other measures for industrial and commercial premises"**

If you are considering modifications to your Terminal to provide tertiary containment, it is worth looking at the ToggleBlok valve. These valves are designed to provide remote isolation of existing underground drain lines.

·······●●●●●●●●●●●●●·······

# CHAPTER 2

# *Terminal Layout*

## 2.1    Hazardous Areas

The definition of a 'Hazardous Area' can change depending on the prevailing National standards. Probably the most commonly used definition is in NFPA 70 which states:

**Zone 0** is one where "ignitable concentrations of flammable gases or vapours are present continuously ... or are present for long periods."

**Zone 1** denotes that, "...ignitable concentrations ... are likely to exist under normal operating conditions," or as a result of leakage or repair operations.

**Zone 2** indicates that ignitable concentrations "are not likely to occur in normal operation, and if they do occur will exist only for a short time."

Areas are deemed hazardous where flammable gas or vapour is present in the atmosphere leading to an increased risk of fire or explosion. Many liquids are capable of producing flammable vapours, so it is not uncommon for specific regions within a tank terminal to be classified as 'hazardous'.

## 2.2    Design Codes

The documents below provide minimum standards for flammable liquids. Clearly, the requirements for non-flammable liquids are much lower:

- Model Code of Safe Practice-Part 2 – Guidance on design, construction and operation of petroleum distribution installations (Institute of Petroleum)

- The Storage of Flammable Liquids in Tanks (UK-HSG 176)

- Safety and Environmental Standards for Fuel Storage Sites (UK-HSE)

- Safety Guidelines and Good Industry Practice for Oil Terminals (UNECE)

- Recommendations on the Design and Operation of Fuel Storage Sites (UK-HSE)

- Directive for Above Ground Storage of Flammable Liquids in Vertical Cylindrical Tanks (PGS 29-Dutch Standard)

In practice, the standards applicable to each Terminal are determined by the products being stored. Tank spacings are set out by NFPA 30 or similar, whilst equipment packages such as water treatment have their standards set either by International bodies or national standards committees.

A good starting point is to look at the internal engineering guides produced by the oil companies. For most Terminals, the requirements are overkill because few of the liquids stored are as flammable as hydrocarbon products, but the guides do give a good overview of the technical aspects to be considered. **GS EP SAF 341** (Total Exploration and Production) and the equivalent **Shell DEP** are typical.

## 2.3    Constraints on Terminal Location

### Air

Airports and airfields tend to have building restrictions at either end of the runways for obvious reasons.

Refueling depots on the airfields are usually located to one side of the main runway and near a parking apron or taxiway. Airports tend to be surrounded by a large open expanse where no buildings exist. Conveniently, cows and wheat do not complain about being next to noisy aircraft, so the land is still in economic use.

Building a bulk liquid storage facility near an airport boundary should not cause any particular concern, but additional questions during the planning application may be raised if the facility is due to hold hydrocarbon products. In the event of a fire, large quantities of smoke may be generated and may impact aircraft movements.

There is one specific case where the existence of an airport in the vicinity will stop the construction of a new Terminal. LNG terminals could, in some extreme circumstances, vent large quantities of gas. This gas, being very cold, will stay at ground level initially but rise as it warms. Any aircraft flying through a cloud of invisible, flammable gas could either have the engines flame out as oxygen is excluded or over-speed as the gas enters the turbine where only air was anticipated. Neither situation is desirable. This is why LNG Terminals are not on the flight paths leading to or from airport runways.

### Sea

It is a reasonable assumption that Terminals located close to ports are there because they want to benefit from large cargo being delivered by ship. However, there are practical constraints on the maximum vessel size.

Ports upstream or on an estuary may have restrictions on the maximum height of the vessel due to bridges crossing the river. Likewise, there may be width restrictions or traffic flow restrictions allowing only one vessel at a time to enter the constricted area (i.e. the piers between spans of the bridge).

More frequently there may be tidal restrictions. Some smaller ports may have sand bars at their entrance, meaning larger vessels can only enter at certain stages of the tide. It is common for vessels outside the ports that have sandbars to have to wait at a safe anchorage until they can enter the port. Some ports also have tidal restrictions at the berth, where a 'pocket' has been dredged for the ship at its berth but access to the 'pocket' may be restricted.

Areas with a high tidal range (such as the North Sea) are also prone to tidal surges and may have flood barriers to protect residents. These barriers may be closed in certain weather conditions, delaying ship movements in and out of the port.

The physical size of the vessel may also present problems. Most ports will require tug assistance for vessels leaving or entering the port (a ship's rudder is ineffective at low speeds). To get the ship around corners in some congested and heavily developed ports (i.e. Antwerp) the tugs may have to use 'turning posts' to swivel the vessel around.

But the biggest constraints on which vessels can be accommodated are the length of the available quay and the depth of water at the mooring.

Vessels moor at the quayside using hawsers (ropes, lines, or cables). Mooring lines extend beyond the ends of the vessel as they have to hold it secure against the effects of wind, current, tide, wave action, swell and surge induced by passing vessels.

In addition, ships do not move sideways easily, even with tug assistance. When deciding the maximum length of the tanker at a berth, the designers assume there will be vessels moored fore and aft of the tanker and assume an approach (or departure) angle of (typically) 5 degrees is necessary to clear those vessels when entering or leaving the tanker berth.

The maximum length of the tanker that can be accommodated is fixed for any particular berth. However, the depth of water at the berth can be increased by dredging. In fact, regular dredging may be necessary to maintain the water depth at the quay.

The following table gives typical dimensions of some 'standard' vessel classes;

| Class | Length | Beam | Draft | Overview |
|---|---|---|---|---|
| Coastal Tanker | 205m | 29m | 16m | Less than 50,000 dwt, mainly used for transportation of refined products |
| Aframax | 245m | 34m | 20m | Approximately 80,000 dwt, which is the AFRA (Average Freight Rate Assessment) standard. |
| Suezmax | 285m | 45m | 23m | Between 125,000 and 180,000 dwt, originally the maximum capacity of the Suez Canal. |
| VLCC | 330m | 55m | 28m | Up to around 320,000 dwt |
| ULCC | 415m | 63m | 35m | Capacity between 320,000 and 550,000 dwt. |

The deadweight tonnage shown above (abbreviated to dwt) is a measure of how much weight a ship can carry. It is the sum of the weights of cargo, fuel, fresh water, ballast water, provisions, passengers, and crew.

Taking all the above into consideration, and the already congested nature of most ports, it is difficult to find space to build a new Terminal with access to a berth of suitable size in an existing port. The only exception seems to be Rotterdam where land is being reclaimed and the port is rapidly extending into the North Sea. If the Rotterdam Port Authority had its way, it will soon be possible to walk from the UK to the Netherlands!

If large vessels cannot be accommodated at quays within the port complex, the alternatives are:

- Construction of a 'finger' jetty. This is typically necessary when Very Large Crude Carriers (VLCC) are to be accommodated:

- Single Point Mooring or Multi-Point Mooring in open water outside the port:

In these cases, the location of the mooring effectively dictates the location of the Terminal. Some of the above items are considered in greater detail in **Chapter 6.3-Marine Terminals** and in **Chapter 2.3-Ship Loading and Unloading**.

## Land

Other issues which may restrict the construction of a new Terminal include;

- Listed Buildings & Monuments;
- Areas of Special Interest;
- Areas of Historical or Social Significance;
- Right of Ways;
- Endangered animals;

### Listed Buildings & Monuments

It is unlikely that anyone would knowingly want to build a Terminal next to a 'listed' (protected) building, not least because land prices would be prohibitive. However, listed buildings come in a variety of shapes and sizes and can be affected by the vibrations associated with the construction of a new Terminal. If your new Terminal is accessed via a road where a listed building is located, there may well be objections to planning consent because of increased traffic (even temporary).

Monuments can be located close to roads or in the middle of a field. If the monument is on the site of a battlefield, forget about moving the monument and putting a Terminal in its place. Monuments to deceased local dignitaries and land owners are usually less contentious, and the current land owners may be amenable to the monuments to minor dignitaries or relations being relocated.

The worst-case scenario is if planning consent has been given for the Terminal but preliminary earthworks unveiled ancient remains or (God forbid) an old graveyard or burial ground.

There are sites in London where new buildings had to be redesigned and constructed to fit around a bit of a Roman mosaic or wall. They are usually only allowed to do this when arrangements are made for the public to have regular access to the relics. London Crossrail was delayed for years when they unearthed a plague burial pit at Liverpool Street and a Black Death cemetery on the site of the proposed Farringdon station.

### Areas of Special Interest

In the US this is a 'catchall' term and generally refers to known archaeological sites, areas of eelgrass, ecological reserves, parks, protected area designations, or any combination of these.

In the UK the equivalent is a **Site of Special Scientific Interest** (known as an SSSI). This is defined as an area that is considered to best represent the natural heritage in flora, fauna, and geology. That definition would suggest that SSSIs are few and far between. On the contrary, in England alone there are 4,000, covering about 7% of the land area. This broad heading includes Special Areas of Conservation and Special Protection Areas.

No doubt other countries have similar protected areas.

It is not just a case of not being able to build within an SSSI. The unwary can create them by accident! Refineries had to have a flare stack for emergency venting and burning off volatile vapours. The area around the flare stack is fenced off to protect personnel from the radiant heat in the event of flaring. With no regular human presence in the area, the wildlife is free to flourish and doesn't seem unduly concerned that they could turn crispy if the flare went off.

Ever keen to burnish their environmental credentials, the oil companies promoted the concept of having a wildlife haven on their doorsteps, inviting schools and local environmental groups along. They were not so impressed when 'Natural England' declared their land an SSSI, which imposed strict limits on what they could do with their land.

## Areas of Historical or Social Significance

With increased interest in the preservation of minority rights and concern for indigenous people's ancestral homelands, this is a potential 'hot potato' item.

Claims for the historical right to land are difficult to prove one way or another, especially when no written records exist and tribes were predominantly nomadic. There were also conflicts and disputes among the tribes, with one tribe displacing another, and then in turn being displaced themselves.

Disputes on land ownership by indigenous people are well documented in America, Canada, and Australia. These sorts of disputes are also found in Africa as a consequence of the Zulu wars in the 1830s, and land ownership in Cyprus following the Turkish invasion in 1974.

## Right of Ways

Within the UK the term Right of Way has a very specific meaning. It is the legal right, established by a grant from a landowner or by long usage, to pass along a specific route through a property belonging to another. A similar right of access also exists on land held by a government, land that is typically called public land, state land, or Crown land.

When one person owns a piece of land that is bordered on all sides by lands owned by others, an **easement** may exist or might be created to initiate a right of way through the bordering land.

A footpath is a right of way that legally may only be used by pedestrians. A bridleway is a right of way that legally may be used only by pedestrians, cyclists, and equestrians, but not by motorised vehicles.

In some countries, especially in Northern Europe, where the freedom to roam has historically taken the form of general public rights, a right of way may not be restricted to specific paths or trails.

If a Terminal is built across the land with a right of way, public access may still be required when the Terminal is in operation. I worked with a Refinery in Egypt where they believed they could get away with re-routing the existing right of way around the site. They were wrong. They ended up with a main public road running straight through the tank farm.

## Endangered animals

The WWF has a formal list of all animals that are endangered, vulnerable, or threatened. Gorillas, rhinos, or orangutans may be a problem at some locations where a Terminal could be built, but it is the smaller and least known animals that cause the biggest disruptions in the planning process in many places.

In the UK, endangered species include bats, dormouse and newts. Everyone involved in a planning application will have similar horror stories to the ones below:

### *Example 1*

*A dead newt was brought in from the road outside the construction site. Panic ensued and two questions needed answering urgently.*

a)   *Was it a 'Great Crested Newt' or just a newt? Great Crested Newts are protected, but the other species are not. Only a specialist can tell the difference.*

b)   *The dead newt was flat. Did it die before being run over by construction traffic, or did it die because it was run over? How do you find someone to do a newt post-mortem?*

### *Example 2*

*To extend and modify a large gas terminal, the planning application had to contain proof that no endangered species would be negatively affected:*

a)   *A specialist 'dormouse detection agency' was employed to prove that no dormouse would be affected. After months of site investigation, they concluded that there were no dormice.*

b)   *There are 18 species of bats in the UK. A group of 'bat hunters' was employed and confirmed we had 14of the 18species living in a building at the end of the construction site.*

*The whole project was canceled by the Terminal Owners before the bat question was resolved. Whether it was a factor in their decision, we will never know.*

## 2.3    Product Movement

### Pipelines

Pipelines are the most efficient way of moving large volumes between locations. Crude oil pipelines are used to move degassed and dewatered oil from the production centre to a port where it can be shipped out on a tanker. Or from a receiving port to a refinery.

It is generally refined hydrocarbon products that are transferred between Terminals. Specialist pipelines for low-volume products exist where there is a continuous demand i.e. a Terminal pumping ethylene to a manufacturer of plastics under a long-term supply contract.

### *Impact on Terminal Layout*

In practice, the existence of a pipeline should have relatively little impact on the layout of a Terminal.

For Terminals pumping into pipelines, pump pits and pump slabs will already be installed and these pumps will frequently have suction manifolds associated with them. The only new plant would be the pipeline pumps which take up little space. The pump discharge has isolating and non-return valves but seldom any other equipment so takes up little space. Pipelines normally run above the ground inside the Terminal, only going below ground as it leaves the site. Before the pipe goes below ground level, there will be an isolating flange (as part of the corrosion protection system).

Pipelines coming into a Terminal are likewise protected by an isolating flange as the pipe rises above ground level (usually just inside the fence line). The discharge from the pipeline can then go straight to the tank inlet valves.

### Single Product Pipelines

The most obvious examples of single-product pipelines are those used to take aviation fuel to major airports and transport hubs.

The dispatching end of the pipeline has buffer storage to even out upstream fluctuations of supply, whilst the receiving location has buffer storage to even out fluctuations in demand. In the pipeline between, the pumps supply a continuous and steady flow of jet fuel (demand for AvGas is too small to justify dedicated pipelines).

Pipelines to airports are a particularly good example of single-product pipelines. At its simplest, it could be a small, single pipeline going from a local terminal to a small regional airport. Pumps can operate continually if a steady flow is required or a variable speed pump can be ramped up and ramped down. For peak flows, multiple pumps, operating in parallel, can be used to satisfy demand.

However, the situation at major international hubs is different. Heathrow in the UK has multiple underground pipelines from multiple Terminals, delivering to onsite and offsite fuel receipt facilities. Other locations have both refineries and terminals pumping into a single pipeline and multiple airports serving a city taking their supplies from the same pipeline.

Flows into and out of the pipeline must be coordinated, and this is done by a central authority

### Multi Product Pipelines

A good example of a large multi-product pipeline is the Rotterdam Rhine Pipeline (RRP), operated by a joint venture of IOCs. One pipeline system transports oil products such as petrol, naphtha, diesel, gas oil and kerosene. Shell (Pernis) and BP refineries in Rotterdam use the pipeline to send multiple products across Holland to Wesseling in Germany. From Wesseling, other private pipelines extend as far south as Frankfurt and Ludwigshaven.

The obvious question is how do you stop the products from mixing and cross-contaminating within the pipeline? The answer is surprisingly simple.

All the products sent down the pipeline are clean, hydrocarbon-based but have different densities. Simplifying the process down for an explanation, you pump a batch of gasoline down the line and shut the pumps down. Now you start up different pumps and open the valves to let naphtha into the line. At the far end of the pipeline, you will be getting uncontaminated gasoline coming through the line, followed by some of the gasoline that is contaminated by naphtha, then a delivery of naphtha that is uncontaminated. The difference between the products is easily measured by checking product density.

By pumping products down the line in small batches, but always in the sequence of petrol, naphtha, diesel, gas oil, and kerosene, you minimise the extent of the contamination between products. The different products in the pipeline do not mix along the pipeline, except at the interface between the two products, and the volume that is contaminated by this mixing is quite small.

In the example above, the gasoline contaminated by naphtha is put into the 'clean' naphtha cargo. Likewise, the contaminated naphtha goes into the 'clean' diesel cargo, and so on down the list of products. In this way, no cargo falls below its product specification.

Clearly, after pumping a batch of kerosene down the line, the following batches will be in the reverse order gas oil, diesel, naphtha, and petrol. Having spoken to Operators on a multi-product pipeline I understand the process is (almost) infallible, and differences, even between supposedly identical cargos of kerosene can be easily identified.

## Construction

Regardless of whether a pipeline is multi-product or not, the actual process of construction is identical.

The first stage of any pipeline is the preliminary engineering and route surveying, closely followed by a period when consents and permits are obtained. This is invariably a long process as way leaves, easements, and rights of ways have to be established legally (1 mile of pipeline in Dubai needed 130 'consents' including from the Dubai Royal Family as it crossed the entrance to their terminal at the airport).

The pipeline right of way is staked out, which shows the extent of the temporary construction area along the route, including storage and staging areas. The width of the right of way is based on the diameter of the pipeline. It has to allow for the pipeline trench, space for the spoil from the trench, and an area for the construction machinery.

Staging areas and storage yards are strategically located along the planned right-of-way. These areas are used to stockpile pipe and to store fuel tanks, sandbags, fencing, stakes, and equipment. They provide parking for construction equipment, employee vehicles, and locations for office cabins.

Staging areas are cleared and covered in gravel, often reinforced with timber matting. These areas can be located in fields, pastures, or forested land and can impact streams and wetlands. Often, these areas require the construction of access roads to and from paved roads, and to and from the areas to the pipeline RoW. At the end of the construction, these areas must be returned to their original condition.

The trench for the pipeline is dug after the RoW is cleared of trees and vegetation. When the trench is completed, pre-coated segments of pipe are transported from stockpiles in the staging area to the RoW. Pipes are strung out above ground beside the trench, and pipe sections are bent to allow the pipeline to follow the planned route and the terrain.

FIG.201 Building pipelines through the desert is a lot easier than through a built-up area, but there are still challenges. Like the owner of the dead camel demanding compensation, who insists it was a pedigree racing camel and pregnant when it stumbled into your open trench.

The pipe sections will then be welded together, sandblasted, and the weld joints coated to prevent corrosion. Finally, the weld joints are inspected with an x-ray to ensure their quality. Connected lengths of pipe can then be lowered into the trench.

Pipelines cross existing roads, highways, streams, rivers, and wetlands. Typically, pipelines are constructed underneath these obstacles by either boring for shallow depth or using horizontal directional drilling (HDD) for deeper placement.

After the pipe is inspected, the trench is filled in and the pipeline integrity is verified with a hydrostatic test. This water is sent through the pipeline and the pressure is increased to above the maximum operational level. If the pipeline remains intact during this test, it is deemed operational. After this, the surface of the ROW is seeded and permanent markers are placed along the pipeline route.

Water is removed from the pipeline and foam "pigs" or scrappers are inserted into the pipe to clean it out. When the pigs eventually exit the far end of the pipe, dry air is blown through the pipeline. The air is sampled and tested for moisture content. When the moisture gets low enough, the complete pipeline is filled with nitrogen.

The longer the pipeline, the higher pressure the pumps have to operate at, and the thicker the wall of the pipe needs to be at the beginning of the route. On long pipelines, it is common to put interim pumping stations along the length of the pipeline. In this way, the maximum pressure in the line can be limited, regardless of the length of the line.

Most readers would be surprised at the extent of buried pipelines in their locality. When you know what to look for, the markers are easy to spot. After a few years, at ground level there is no sign that the ground was ever disturbed.

### In-Line Inspection

Pipelines are protected by an impressed current cathodic protection system. Checking to see if the corrosion protection is working correctly is the work of specialist inspectors.

With buried pipelines, external corrosion is a major risk, but internal corrosion is also a significant concern. It is necessary to inspect the line regularly to ensure its integrity, but digging up the pipeline for external inspection is not an option. The solution is to inspect the pipeline from the inside to detect faults and metal loss on both the outside and inside surfaces of the pipe.

Intelligent pigs are used to inspect the pipeline with sensors and record the data for later analysis. Surface pitting and corrosion, as well as cracks and weld defects, are often detected using "magnetic flux leakage" pigs. Other smart pigs use acoustic transducers to detect pipe defects. Caliper pigs (also known as gauging pigs) can measure the roundness of the pipeline to determine areas of crushing or other deformations.

During the pigging run, the pig is unable to directly communicate with the outside world due to the materials that the pipe is made of. It is therefore necessary that the pig use internal means to record its movement during the trip. This may be done byodometers or gyroscope-assisted tilt sensors.

Location verification is often accomplished by surface instruments that record the pig's passage by either audible, magnetic, radio transmission, or other means. The sensors record when they detect the passage of the pig. This is then compared to the internal record for verification or adjustment.

Like all inspection techniques, there are continuous improvements in technology and methodology. There are plenty of companies offering contactless pipeline route inspections by helicopter and drone, including magnetometry. It is still too early to say whether these new methods will supplant the use of intelligent pigs. However, there is one area where they may succeed and that is with un-piggable lines. These have been created where there are multi-diameter lines, lines that have branches or sharp bends. However, the technology is proprietary and little information is currently available.

## Road Tankers

### *Overview*

In most situations, road transport provides the most flexible delivery method. Every developing nation has high-speed roads between the major population centres, and the network of roads is expanding all the time, even if maintenance standards are variable. Where the minor road network is limited or of a poor standard, delivery vehicles are adapted to cope with the dust, heat, rutted surfaces and mud.

In Europe, articulated (5th wheel) tractors and trailers with a capacity of 44m3 are frequently used to deliver gasoline and diesel to filling stations. Where road conditions are more of a problem, smaller non-articulated tankers are used. In the worst case, robust rigid chassis tankers carrying 10 or 15m$^3$ are the popular choice.

### *Advantages*

- *Road transport is the cheapest means of transport available;*
- *Road transport can adapt to delivery times and quantities;*
- *Road transport has no limitations on the shipment of flammable or toxic products;*
- *GPS and fleet management software make it easy to always locate and track deliveries;*
- *Road transport provides a door-to-door delivery service;*

### *Disadvantages*

- *Accident rate!*
- *Vehicles have limited load capacity;*
- *Capacity can be reduced by poor roads and infrastructure;*
- *Traffic jams, especially on access to cities and industrial estates, can lead to delays;*

FIG.202   Even four loading bays can be accommodated on this Terminal's perimeter road. The design problems become more onerous when the frequency of loading increases.

### Impact on Terminal Layout

The impact on the Terminal layout is usually a function of the number of road tanker movements.

Consider a mixed product Terminal where a tank pit contains a product (or range of products) that needs to be exported by truck. If the loads are small, or the tanker loadings are infrequent, the loading bay can be located on the Terminal perimeter road with minimal impact on the layout of the tanks or pits. Remember that most Terminals will have a mandatory distance between the tanks (or the tank pit) and the fence line.

A typical single vehicle loading bay has an overall width of about 7m, so can easily be accommodated alongside the Terminal perimeter road, on land between the fence and the road that would normally be unused. If more than one vehicle is to be loaded at the same time, a loading 'island' with a tanker loading bay on each side requires about 9m.

### Flammable Liquids

Where flammable liquids are concerned it may not be so straightforward. International oil companies tend to have their own internal standards for the layout of vehicle loading facilities. Shell has a DEP (Design and Engineering Practice 31.06.11.11) that is very specific. Amongst other things, the DEPs stipulate distances from the fence line as well as standard road geometry and widths and loading bay dimensions. The DEPs are proprietary, but most companies follow similar guidelines.

### Improving Operational Efficiency

The loading bays pictured above will be suitable when the number of road tankers being loaded is small. If vehicle loading becomes more frequent, the number of loading bays needs to increase and the turnaround time needs to reduce.

Terminal Owners need to reduce their overheads to remain competitive. This must be done without compromising safety or security. Even without physical changes to the equipment, changes in operating methods can have a considerable impact on throughput.

This may involve:

- Increasing tanker size;
- Changes to shift patterns, including staggered starts and pre-loading, up to double shift or 24-hour working;
- Improvements to traffic marshaling;
- Reducing peak demand to spread the workload across the working day;
- Reducing vehicle idle time;
- Loading with multiple arms simultaneously;

One could even resort to increasing the pump capacity to shorten loading time, but the practical limit is 2000 litres/min. at each arm due to the risk of static electricity as the liquid swirls within the tank.

When designing the layout, vehicle loading bays are assumed to have their 'peak period' first thing in the morning.

This is because most tanker deliveries are organized for the tanker to go out full first thing and return empty at the end of the day. Over the rest of the day, vehicle arrivals will essentially be random, occupancy will fall and average waiting times will reduce.

FIG.203 A modern road tanker loading bay. Access to the top of the tanker to open the hatches is via a gantry that swings into place. Even then, Operators wear a safety harness in case of a fall.

### Vehicle Maneuvering

At times of peak demand, there will be vehicles waiting for a loading bay. Therefore inside the Terminal there will be parking for waiting vehicles. In practice, this waiting area does not need to be within the Terminal-having tankers wait across the street works just as well!

A typical road tanker can be up to 14 meters long and even articulated vehicles have a large turning circle. Therefore the amount of maneuvering the tanker has to do within the Terminal, the more physical space is required.

A typical large road tanker loading operation, such as deliveries for ground fuels, will require:

- Access from a surfaced road with good connections. It is preferred that the tankers can turn off the road without crossing oncoming traffic, even better if there is a slip lane;
- Space to stop the tanker in front of security gates;
- Space for some tankers to park while awaiting a free loading bay;
- Space to turn the road tanker into the loading bay;
- The loading bay;
- Space to turn the road tanker when leaving the loading bay;
- Space for some tankers to park while the drivers collect Bills of Lading etc. In older Terminals the Terminal Operators issue these documents which means the tanker drivers have to park up and go to the Operator's office to collect them;
- Access to a surfaced road with good connections. Tankers needing to cross traffic on the main road should be controlled (by traffic lights or similar);

### Security

In all Terminals, and especially those that have to comply with ISPS requirements, all vehicles and drivers are security checked before they are allowed on site. However, many of these drivers will be regular visitors to the Terminal, so security checks can be somewhat cursory. It may also be a factor that many of the products stored at the Terminals have little value to thieves. Quite what a thief would do with 10,000 litres of concentrated detergent I don't know, but 10,000 litres of street grade gasoline would find ready buyers.

In multi-product Terminals that also handle gasoline and diesel, the road tanker loading area is frequently regarded as a separate security zone and is fenced off from the remainder of the Terminal. Drivers accessing the tanker loading facilities do not have security clearance to access the rest of the site. This is not surprising as drivers and vehicles change all the time, especially when 3rd party transport companies are contracted to provide vehicles and personnel.

### *Loading Operations*

In a small depot, it is common for the tanker loading equipment to be operated by the Depot's employees as the tanker drivers tend to be of variable quality. Also, the loading process is more manual and therefore more open to mistakes anderrors. As soon as the tanker is parked in the loading bay the driver is sent off to wait in a secure area. This also allows drivers the opportunity to refresh themselves between journeys.

In the larger Terminals that can justify the investment to reduce labour costs, the truck loading facilities are automated to reduce turnaround time and decrease the manual content of the operation. Here the drivers are responsible for loading their own vehicles. Apart from reducing the number of Operators the Terminal has to employ, the work of HGV drivers is regarded as skilled and well paid.

In the UK, most drivers have a Class 1 HGV license as well as additional training to carry dangerous goods (the ADR treaty). They are therefore considered competent to load and unload flammable liquids, gases, oxidising agents as well as toxic and corrosive materials.

Some of the above topics are explored in greater detail in **Chapter 6.2-Ground Fuels**. This section also contains a detailed explanation of the vehicle loading process, safety issues and operations.

## Rail Tankers

## *Overview*

Rail serves two distinct markets;

- Terminals import product in bulk by ship or barge, store it until wanted by the Customer and then forward the product by rail. This usually involves frequent, regular deliveries of consistent parcel size. In this case, the Terminal will have the facilities to **load** the rail wagons;

- Trains bringthe refined products from the Refinery or the port to the distribution Terminals. In this case, the Terminal will have the facilities to **unload** the rail wagons;

Unfortunately, despite the obvious environmental advantages of using the rail network, practical constraints mean that movement by rail is a declining market. I offer two examples:

### *Example 1*

*Until 1991, Ukraine had been a Soviet Republic and was part of the extensive rail network covering the old Soviet Union. The fact it was broad gauge was a minor inconvenience.. After all, Ukraine could refine Russian crude and what wasn't refined within the country could be imported from Russia.*

*By 2022, all Ukraine's refineries had been shut and there were no imports of fuel from Russia due to an embargo. Shipments directly by rail from Europe were not possible, broad gauge not being compatible with European rolling stock.*

*The Ukrainian depots continued to deliver to their customers. It was done using road tankers, delivering to petrol filling stations, schools, hospitals, agricultural users, and utilities. Whereas rail tankers have one route, road tankers have the flexibility to alter routes, alter the timing, change delivery quantities and vary the product mix, all at short notice.*

## Example 2

*A Terminal in Rotterdam has a dedicated rail loading facility. I was told it wasn't in use and there was little hope of it ever being used again. It had been built for a Customer in Switzerland but the contract had been terminated.*

*Operators on the site said there were always problems getting the right rolling stock delivered to Rotterdam when they needed it, and always delays waiting for the wagons to be collected by the local rail operating company. Deliveries to the Customer usually took 7 days from dispatch but were erratic. In the end, the Customer changed to a local supplier who could guarantee regular deliveries.*

That is not to say that rail facilities should be avoided when planning a Terminal layout. Rail tanker sizes up to 70m³ per wagon give clear financial advantages over road transport and obvious environmental benefits. It is more a case of weighing up the pros and cons and being sure product movements by rail fit into the business model and future plans of the Terminal. In particular, do the intended Customers and future Customers have rail access with connections to the main line and the space/manpower to deal with the rolling stock?

### Advantages

- *A 'greener' option as trains burn less fuel per ton-mile than road vehicles;*
- *Rail is 'safer';*
- *Train tankers carry more product than road tankers (up to 2.5 times the capacity of a large road tanker);*
- *Long-distance freight movement is cheaper and quicker by rail;*
- *Cargo does not need to be offloaded multiple times as sometimes happens with road deliveries*

### Disadvantages

- *Customer needs similar investment in rail facilities;*
- *Not economical over short distances. Road transport is quicker and more flexible;*
- *Road transport can make door-to-door deliveries, including 'last mile' services;*
- *Road transport is easier to track cargo and has better security;*
- *Rail can get delayed at border crossings;*

## Containerised Loads

A recent buzzword in the freight industry is "intermodal" (or multimodal) although this doesn't seem to have filtered down to the bulk liquid storage Terminals yet.

The logic is that intermodal transport combines the best of rail and road transport systems. Since their introduction in the 1960s, standardised containers have moved around the world by ship and train. When the train gets to the terminus, the container is loaded onto a lorry to complete the final part of the journey to the customer's yard. Now there is a 'liquid' equivalent to the steel container.

ISO tank containers are designed to transport hazardous and non-hazardous liquids in bulk. Tanks are cost-effective and reduce shipping and handling costs. The tank is a cylindrical-shaped container, primarily made of stainless steel. They are insulated and protected from corrosion and harsh weather conditions.

ISO tank containers come in 20ft and 40ft dimensions. However, the 20ft tank is the most common and has a capacity of up to 26m³. Using ISO tank containers, a Terminal supplying specialist liquids can use a combination of rail (for long-distance travel using flatbed rail wagons) and road transport (for 'last mile' delivery to the Customer).

### *Impact on Terminal Layout*

The adoption of rail transport is fundamental to the choice of a Terminal's location, as well as the layout of facilities inside the perimeter fence.

#### *Access to the Mainline*

A Terminal that will rely on movements of cargo by rail has to be near an existing (or planned) main line. Extending a mainline or branch line to a new location, unless the rail network has financial guarantees, will be difficult. When planning the location, the layout of the lines is critical as it is necessary to decide whether it will connect to through track or other sidings. Sidings can be constructed using steel rails of lighter weight which are meant for lower speed or less heavy traffic. Sidings do not generally need signals. Sidings connected at both ends to a running line are known as loops. Those sidings not connected at both ends are known as dead-end sidings.

Rail lines are designed in the same way as pipeline routes and roads. The alignment of the track must take into account the slope of the ground, the radius of the bends, and the maximum length of the train on the sidings.

Inside a simple fuel depot, a single dead-end spur from the main line is all that is needed. A loco, pushing the rail tankers destined for the Terminal moves up the spur until the wagons and the loco are clear of the main line. At this point, the loco uncouples and goes off to do other work.

The spur line will generally have rail tanker loading or unloading facilities at 2 or 3 positions along the track. As the tanks are filled or emptied, the wagons are moved forward until the next wagons are aligned. The wagons are moved forward by a pulley system to avoid the need for shunting locos. It is not uncommon for unloaded wagons, as well as wagons full of product, to be sitting outside the Terminal on the spur line, either waiting to be unloaded or for the empty tankers to be collected.

The spur can either be privately owned or owned by a railway company. The locos and drivers doing the shunting are usually provided by a railway operating company.

When loading or unloading has finished, the wagons will be moved to a nearby siding, awaiting the availability of similar wagons before being retrieved by the Rail Operator.

#### *Rail Tankers*

There is a significant variation in rail tank sizes, ranging typically from 50m3 to 100m3 in Europe, although some larger tankers are available. Tankers in the US, such as the DOT-111 type, are larger as they have different restrictions on the distance between the bogies and the number of wheels. Tankers designed for use with heated products can be supplied with insulation and hot water heating pipes.

Where a particular style and size of rail tanker can be consistently delivered by the railway operator, more loading/unloading positions can be installed to reduce the overall loading times.

### Combined Road/Rail Loading Facility

A practical and cost effective option is to have a combined road and rail tanker loading bay. The products being loaded into the train and the road tanker may be the same, which means all the product pumps and lines are the same. At the same time, the infrastructure (fire, spillage, metering) does not need to be duplicated and the rain shelter provides weather protection regardless of which type of tank is being filled.

These types of facility are generally restricted to 2 loading bays, one on either side of a single 'island' and are associated with the products in a particular tank pit. Therefore it is not uncommon to have a number of these combined loading bays around the site, each associated with a specific tank pit, product or Customer.

Where loading bays are associated with specific tank pits they can be built adjacent to it, reducing the quantities of pipework involved. They can also be tucked away in odd corners of the Terminal that are too small for additional tanks, as long as the rail track and the road can be aligned to suit.

### Loading Product

The impact on the Terminal layout can be significant due to the space taken up by the track, the loading facilities, and all the associated sidings. A typical configuration would have 2 dead-end branch lines from the main line with buffers at the end of the branches.

Parallel to the branch lines, there are frequently several sidings. The point of the sidings is to receive empty wagons when they become available, carry out any cleaning necessary, and store filled wagons until sufficient is available to justify bringing a loco on site to remove them. It is not uncommon for large sites to have their own shunting locos to move wagons between sidings.

When loading the rail tankers, the process is the same as with road tankers. Although top loading was originally used, it is now the norm to 'bottom load' products, using small mechanical loading arms. Some liquids may produce VOCs or greenhouse gases and vapour recovery hoods may be placed over the open hatches of the tanker.

FIG.204   The space to load one or two rail wagons isn't great but the practical constraints of track routing and geometry from the mainline and within the Terminal are a problem.

FIG. 205   A busy rail tanker loading centre for chemical products. The shed in the background covers the loading bays and vapour recovery equipment. Wagons in the foreground are empty, being cleaned or awaiting dispatch.
Photo; Koole Rotterdam

Many loading facilities are roofed, allowing loading to proceed under any weather conditions. The risks to Operators working on the top of rail tankers are just the same as working on road tankers, so access platforms are generally installed to make opening the top hatches and placing of the hood safer.

Many loading facilities are roofed, allowing loading to proceed under any weather conditions. The risks to Operators working on the top of rail tankers are just the same as working on road tankers, so access platforms are generally installed to make opening the top hatches and placing of the hood safer.

### *Unloading Product*

The equipment necessary to unload rail tankers is straightforward.

The wagons are unloaded through the bottom connection, and as it is only gravity feed and not pressurised, flexible hoses are frequently used.

When mechanical loading arms are used, they are generally simple and robust. Ideally, they should have an extended reach to make them suitable where the wagon can not be accurately spotted. Arms can be counterbalanced to make handling easier or can have spring cylinders to assist the Operator in connecting to the tanker. Swivel joints and mechanical seals allow the arms to move without leaks with minimal maintenance.

Access to the top of the tanker is still required for Operators to crack open the tank hatches but vapour recovery equipment is not required on unloading. Access to the top of the tankers is usually facilitated by Operator platforms, which may incorporate safety handrails and fall protection devices.

There are no level indicators on the tankers, so unloading has to be supervised. When the tank is empty, the loading arm is disconnected and pulled back out of the way and the train moved forward.

FIG.206    Rail tanker unloading equipment at a small fuel distribution depot in eastern Europe. A single-track, dead-end line serves only the depot. Deliveries are weekly during the summer when demand is at its peak.

FIG. 207 Unloading diesel. As each wagon is emptied the train is moved along the track by an electric winch and cable.
It does not get any simpler than this.

\* \* \*

## Ship Loading/Unloading

### *Marine Facilities*

Terminals with a sea/river frontage will usually have provision to berth ships or barges.

Berths come in 2 broad categories:

- A finger jetty-which extends out into the river or sea at about $90^0$ to the shoreline;

- A quay or wharf-which runs parallel to the shoreline;

The primary purpose of these two structures is to provide an area the ships can lay alongside to load/offload, whilst providing secure and robust moorings. If you look at a large modern port, berths can be a variety of arrangements, and it is not uncommon to find berths that combineelements of both.

### *Jetty*

A jetty is constructed when there is a shallow sloping shoreline. Having a jetty extending out from the shore allows vessels with bigger draughts to be berthed without the expense of dredging. Jetties rest on concrete or steel piles, driven into the river/sea bed. A typical jetty at a fuel terminal will consist of 3 main areas;

- **Trestle** leading from the shoreline to the jetty head. Sometimes this trestle is wide and strong to allow wheeled vehicles access to the jetty head, but more frequently it only allows foot access for Operators. Running along the trestle are the pipelines connecting the tanks on shore and the loading equipment on the jetty head.
- The **jetty head** crosses the end of the trestle and provides a working space for the ship's loading equipment and services. The front face of the jetty head is frequently timbered (fenders) to minimise impact damage to vessels when they berth. Alternatively, there may be large rubber pneumatic fenders (Yokohama type). As vessels are berthed, the side of the ship 'leans' on the fendering along the front of the jetty head so the piling arrangement is designed to allow some flex in the structure;
- Either side of the jetty head (typically upstream and downstream) are two or more **dolphins**. These are piled concrete platforms that allow very long vessels to use the jetty. They can be connected to the jetty head by sliding gangways or freestanding (in which case the mooring crew arrives by service boat). The dolphins are there to take the loads from the vessel's mooring lines. At each dolphin, there may be a powered capstan and quick-release hook system but some dolphins may just have a fixed bollard for the ship's mooring line.

### *Quay*

A quay (pronounced 'key' and derived from a French word) permits more flexible use.

Whilst a jetty is designed specifically for the loading and unloading of liquid cargos, quays can be used for containers, general cargo, bulk dry materials (such as coal), and liquids. There are no dolphins and all the mooring bollards are built into the quay. Because of this, it is unusual for VLCC (very large crude carriers) to be accommodated at quays.

The quay does not flex so Yokohama fenders are used to protect the side of the vessel from rubbing the concrete quay. The area in front of the quay is usually dredged.

### Berthing Vessels

Most ports insist that vessels have tug assistance when approaching or leaving a quay or a jetty. At slow speed, the ship's rudder gives little control over direction. Even using the ship's thrusters doesn't give precise enough control without the tug's assistance.

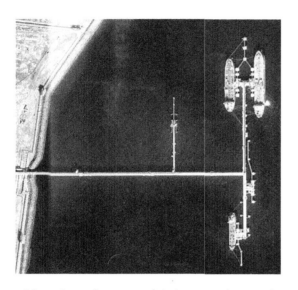

FIG. 208    A jetty at Fujairah, UAE. The trestle back to shore, the jetty head, and the dolphins are all visible.
Photo; Google Earth

FIG. 209    A typical quay. The vessel shown is used for general cargo. If a liquids tanker was berthed here it would be unloaded using a movable loading arm.

Ports employ contracting companies to provide the shoreside personnel to carry out the manual work associated with mooring a vessel. They take mooring lines (also known as hawsers) over to the dolphins, where they are slipped over the hooks. The winch/capstan on the ship then tightens the mooring lines, leaving enough slack to allow for the ship to sit unconstrained at the berth.

On most large vessels there will be at least 6 mooring lines. Head and stern lines extend beyond the ship's stem and stern. Breast lines keep the ship close to the quay or jetty. Head and aft springs stop the vessel from moving backward and forwards with the currents. Modern practice is to monitor mooring line stresses with alarms on board the ship.

### Constraints

When considering what ships can use the berth, the governing factors are:

- Overall length;
- Beam (the breadth of the ship at the broadest point);
- Draft or draught (the vertical distance between the waterline and the bottom of the hull or keel)
- Deadweight (usually measured in tons);

For Terminals (or Refineries) importing crude oil, the bigger the vessel the lower the transport costs. So the most popular vessels for crude cargos are VLCC. ULCC are still fairly rare so few places are equipped to take vessels of this size.

Tankers that carry refined products or liquid chemicals tend to be smaller. The refined product is being distributed so the tankers use smaller ports. Most cargos are also multi-product, so having 10 or more tanks, each containing a different product is common. Specialist chemical coasters can be between 5,000 and 35,000 DWT-refined product tankers up to 100,000 DWT. But the range of vessels is very wide and covers everything from ships that carry only orange juice to ships that carry specialist liquefied gases.

There are extensive regulations that relate to dangerous cargos, such as MARPOL and SOLAS, but that is a specialist topic and well outside the remit of a 'primer'.

**"International Safety Guide for Oil Tankers and Terminals"** gives extensive technical information about the berth, the ship connections, and firefighting and fire protection. This is recommended reading for anyone involved with the design and operation of a marine terminal.

### *Product Loading/Unloading-Ships*

A marine loading arm is frequently installed at the jetty head or on the quayside to load and offload liquid cargos from ships.

A marine loading arm, also known as a MLA is a mechanical arm consisting of articulated steel pipes that connect a ship to the on-shore equipment. Generic trademarks such as Chiksan (often misspelled Chicksan) are often used to refer to marine loading arms.

The advantage of a loading arm is that it can operate at higher pressures and have higher flow rates than hoses. When ships are chartered there is an agreement between the parties on the time the ship will be tied up. Any time beyond this and the Ship Owner has to be compensated by the Charterer with a fee called 'demurrage'.

An ideal arrangement is to turn the ship around in a day (nominally 25 hours for 2 tides). Therefore, when the loading system is designed it is usually optimized to offload a full cargo from a typical vessel in a day. Larger ships such as VLCCs may have turnaround times of 36 or 48 hours.

Whilst large diameter loading arms will offload quickly, smaller arms offer greater flexibility, higher availability, and reliability.

Although loading arms can be cleaned, it is usual to have dedicated arms for 'white' products and 'black' products. Where heated products are being transferred, the loading arms can be supplied electrically trace heated and insulated.

Originally, loading arms connected to the ship's manifold using a bolted flange. A more recent configuration has been quick release 'no spill' couplings as a precaution against a ship breaking free of its mooring lines. Many ports insist the whole product transfer operation is supervised by staff monitoring for leaks etc. That practice will likely continue as the fallout following a leakinto port water is significant for all involved.

Once connected, the hydraulics are disconnected and the arm is allowed to move freely. As the arm uses swivel joints, it can follow the movement of a moored vessel at its berth. Small loading arms can be manually operated and can even be supplied on trolleys, making them movable on the quayside.

The OCIMF specification for loading arms for typical vessels allows 5 metre sway from the berthing line. Ship movements up and down are called 'heave', 'surge' covers movement fore and aft, whilst 'roll' describes movement around the centre of buoyancy.

A typical arrangement at a medium size terminal would be two 6" or 8" arms for white products and a similar arrangement for black products. 16" arms would generally be used for larger vessels such as VLCCs. The arms must be located on the quay or jetty head to avoid their counterbalance weights clashing.

For people not accustomed to extreme tidal ranges, it can be surprising the changes to the vessel freeboard during a loading operation. Areas around the UK and Northern Europe have a normal tidal range of 5-6m, but spring tides, neap tides, and tidal surges can make the effective tidal range even greater. Ports in the Caribbean have negligible differences between high tide and low tide whilst Accra (Ghana) has a tidal range of 1 metre. The loading arms have to be designed to reach the manifold of the smallest (laden) ship at low tide, as well as the largest (unladen) ship at high tide.

A loading arm must be drained or closed off before the connection is broken off. For fuels such as gas oil and diesel, the lines can be blown out with high-pressure air. In the case of more flammable fuels such as kerosene or petrol, the loading arms are usually 'stripped' with pumps.

For ships being loaded, a vapour recovery system is important as large volumes of vapour will be displaced as the ship's tanks are filled. Vapour recovery lines are not subject to the pressures in loading arms so hoses are sufficient. The flexible hoses are generally stored on a hose reel, and supported by a crane during loading.

The most exotic of marine loading arms are those used for handling LNG. As LNG vessels carry cargo between 125,000 and 260,000m³, large diameter arms are needed. Even then, LNG tankers may take 3 days to load or unload using three arms (2 LNG and a vapour return). For safety, arms are equipped with automatic shut-off and disconnect valves. In the event of an emergency, the arms disconnect and the vessel can sail away. The additional weight of the equipment at the end of the loading arm means that the supporting structure must be particularly robust.

### Product Loading/Unloading-Barges

Many people will be familiar with the flat bottomed barges pushed or pulled along our inland waterways, but the powered barges used on Europe's major rivers are shallow draught versions of coastal tankers. As far as the equipment to load and unload them, they are little different from their bigger cousins.

Hoses still have a valuable place for loading and unloading smaller ships and barges. The old heavy rubberised hoses have now been replaced by lightweight hoses that are easier to handle.

Hoses for loading barges are usually hung from a steel gantry on the quayside. The weight of the hose is compensated for by spring supports. When not in use they are allowed to drain into a catchpit at the bottom of the gantry.

There is a hybrid arrangement which is a cross between a marine loading arm but with the outboard arm replaced by a hose. The advantages of this design are not obvious and few of them have been installed.

## Offshore Options

Where there are no deep water berths in the port, there are still two options for offloading liquid cargos from tankers.

### Single Point Mooring

Single point mooring (SPM) is a floating buoy anchored offshore to allow the handling of liquid cargo such as petroleum products from tankers. Located several kms offshore from the on-shore facility and connected using sub-sea pipelines, a single-point mooring (SPM) can even handle vessels of massive capacity such as VLCCs.

The SPM is moored to the seabed using suction piles, anchors, anchor chains, chains, etc. The SPM consists of a buoy, anchored to the seabed but with a central turret that can rotate. The buoy connects to the tanker using a hawser. The mooring arrangement is such that it permits the tanker to rotate freely around the SPM as required by wind, waves, and current (weathervane).

The product transfer system is located at the heart of the mooring buoy. There is a subsea pipeline between the shore and the Pipeline End Manifold (PLEM) on the seabed. The PLEM connects to a flexible hose (the riser) which is connected to the swivel on the bottom of the buoy. The buoy is connected to the tankers using floating hose strings, which are provided with breakaway couplings to prevent oil spills.

### Multi-Buoy Mooring

Similar in many ways to the SPM, a multi-buoy arrangement uses several buoys fixed to the seabed. Whilst the tanker connected to the SPM will swivel around the anchor, a ship moored at a multi-buoy mooring does not.

The buoys are fixed to the seabed using anchors in a rectangular pattern (two buoys at the bow, two buoys at the stern) that allows the safe mooring of a vessel that is positioned between them.

In this method, the bow of the ship is secured by using both her port and starboard forward anchor lines connected to the mooring buoys at her bow. To avoid swinging at anchor on a change of tide, the stern of the tanker is secured to the mooring buoys at her stern. This mooring holds the vessel in a fixed position and does not allow it to weathervane. As soon as a tanker is approaching the mooring pattern with the aid of tugs, the mooring crew take the tanker hawsers one at a time and tow them to the various mooring buoys.

When the vessel is secure, a flexible hose is connected between the tanker's outlet manifold and a buoy at the side of the ship. The buoy connects to the PLEM and subsea pipeline in the same way that an SPM does.

For an SPM, ships have flexibility over which direction they approach from, and connection to the mooring buoy can be achieved in 15 minutes. Multi-buoy arrangements have more limited approach angles and take longer to secure at their moorings. When the ship does not weathervane, the area occupied by the moored ship/anchors is less. Also smaller vessels can come alongside and take cargo on board for onward distribution to other ports.

For both these arrangements, vessels up to 80,000DWT can be accommodated in water up to 100m deep. The choice is determined by the most suitable anchor arrangements.

## 2.4    Common Services

### Fresh Water

Small depots in remote locations will frequently have no fresh (drinking) water supply. Operators survive perfectly well using bottled water or water sourced from the nearest village. To a large extent, the demand for potable water is a function of the number of people on the site.

If there is a supply of potable water available, the Terminal will usually have a stop cock (pressure regulator and valve) and a flow meter in a pit somewhere near the fence line. If a big enough connection is available (and there is sufficient pressure at the main), fresh water is also used for:

- Catering;
- Ablutions including washing & showers;
- Supplies to WC;
- Emergency showers and eyewashes;
- Supply to Fire Water Storage Tank;

On some more modern Terminals, there may be a system for collecting and using 'grey' water, where rainwater and 'clean' wastewater are collected on site and reused for tasks such as toilet flushing and irrigation.

### Sanitary Drain

Small depots in remote locations will frequently have no flushing toilets. Operators survive perfectly well using traditional 'long drop' toilets. Some locations may have a septic tank and a soak-away. To a large extent, the demand for flushing toilets is a function of the number of people on the site and the availability of service water.

When Terminals are in built-up areas, it is common to have a sanitary waste system on the site that ties into the public sewer. However, it is not unknown for larger Terminals, or those in more remote locations to have their own sewage treatment works. This does need careful design and operation as sewage treatment works best when there is a steady and consistent flow of raw sewage.

As it is not practical to rely on gravity flow of sewerage over large distances, all the waste systems in a location will flow to a central sewerage pump chamber. From here the waste is pumped away to join other outflows or directly to the public sewer.

### Storm water Drain

In Depots and Terminals that are paved, rainwater may be a problem.

Underground drainage systems on some Terminals cannot be used as heavier than air fumes and vapours can collect in the drain and create a hazard. A commonly used solution is the use of pre-cast concrete box sections or Vee profile open top culverts. These are installed to one side of the access roads in 'clean' non-hazardous areas, allowing rainwater falling on the concrete or asphalt surfaces to drain off.

The topography of the site dictates the drain layout and collection points. At the end of a culvert run, there may be either an attenuation chamber or a section of an underground culvert, or a concrete manhole with an integrated pump. The culverts are laid to fall by not less than 1:200 to either a collecting area or a retention pond. The purpose of the retention pond is to collect rainwater during periods of peak rainfall, so that discharge to the environment can be regulated.

Rainwater collected from roofs of buildings and road surfaces is regarded as 'clean' but routine testing of samples should take place before disposal off-site. This is to ensure that mandatory pollution limits on water discharges are not exceeded.

## Oil Contaminated Water

Water drained from the tank secondary containment or from the external floating roof drains is regarded as potentially contaminated.

At many sites, if the water is contaminated with hydrocarbons, it goes to an oily water separator. The most basic version is known as an API separator. Over the years there have been many improvements and developments to the basic concept, resulting in a selection of efficient proprietary systems now being available.

The API separator is a gravity separation device that separates the oil from the water and any suspended solids. Contaminated water flows into a concrete or steel open-top box slowly and steadily. Trash screens remove floating debris and the contaminated water flows into the main body of the separator. Lighter than water oil floats to the surface whilst heavier than water suspended solids collect in the bottom of the separator. The cleaned water sits between the two extremes and flows over a weir to an outfall chamber. Oil sitting on top of the water accumulates and is removed using a slotted pipe as a skimmer. It can either be disposed of off-site (to a certified disposal facility) or can go for further treatment and cleaning. The bottom sediment is removed by a sludge pump.

Similar oily water separators are made by many specialist water treatment firms that use inclined plate packs (coalescing plate separators) to improve the efficiency of the separation of the oil from the water.

No matter how good the oily water separator is, it still has practical limits and in most cases can be regarded as pre-treatment only. They do not work on heavier than water hydrocarbons and do nothing to address pollution by chemical products.

Depending on the source of the contamination, there are three principal methods of removing pollutants:

• Anaerobic – breaks down organic contaminants using anaerobic microorganisms;
• Aerobic – breaks down organic contaminants using oxygen and aerobic microorganisms;
• **Dissolved Air Flotation** (DAF) – separates suspended solids, oils, and greases;

The type of unit most frequently seen in Terminals is the DAF. They work by dissolving air in the water under pressure and then releasing the pressure in a flotation tank. This removes the suspended matter such as oil or solids. When used with chemical pre-treatments such as coagulant, pH adjustment, and flocculation, the clean wastewater produced can be safely discharged into the environment.

## Fuel Gas

Some Terminals need a supply of gas for heating products that become 'frozen' at ambient temperatures. In some cases, the gas can be derived from a process on site, but the majority rely on supplies of natural gas piped in from their national Grid.

The use of 'duel fuel' heaters is not uncommon where the supply of gas is likely to be interrupted-gas oil or diesel being the preferred alternative. In most cases, however, no alternative is provided. The gas supplies are regarded as robust and are seldom interrupted for long. In addition, the tanks have a lot of thermal inertia and can take a long time to cool to a point where there is any real concern.

## Service Gas

Many larger Terminals use nitrogen. It is stored as a liquid but used as a gas.

As noted elsewhere, nitrogen can be used as an inert gas on top of the product in the tanks. In these cases, it may be used to 'blanket' the product to prevent deterioration and evaporation loss. Alternatively, injecting nitrogen into the headspace will reduce flammable vapours below the LEL to improve safety.

It can also be used to commission gasoline pipelines. Nitrogen is injected into the line to purge the air and then gasoline is introduced, expelling the nitrogen as the line is filled. This avoids an explosive atmosphere developing inside the pipe.

Operators can use nitrogen to blow down product lines when there is a need to empty them.

The cheapest source of nitrogen is to buy it by the tanker load and store it on-site in bulk if demand justifies the volumes. Smaller users can buy their nitrogen by the bottle. It is readily available in most countries.

A convenient option is for the Terminal to generate its own nitrogen on-site. In that way, there is no reason to store more than is required for immediate use. As the nitrogen is extracted from compressed air, there are no consumables or waste, and only electric power is required.

## 2.5    Shared Services

Although it occurs comparatively infrequently, sharing resources with your neighbors does have many advantages.

The problem is that all parties must have common needs, even if they are not in the same line of business. Probably the most frequently seen is the sharing of a ship berth.

For most Terminals, ship movements are an infrequent occurrence and it makes sense to share access to the berth with other users of the port. Ship loading arms do not take up a lot of space and can easily work alongside other handling machinery, or be moved out of the way temporarily. Frequently, bulk liquid storage facilities will be adjacent to break bulk Terminals, handling products such as iron ore, coal, and other minerals. These products need conveyors to move the cargo to the berth and onto the ship and the storage areas have to be located immediately behind the berth. Liquid products only need pipelines to transfer the cargo, so these Terminals can be more remote from the berth.

Where a nation has constructed a deep-water liquid import facility as part of its infrastructure development, it will pay for itself quickly when utilization is high. This will only be possible when all local commercial companies are using it.

A simple example is the port of Accra in Ghana where tankers are offloaded at a multi-buoy mooring outside the port and products are transferred to the shore by a subsea pipeline. Around Accra, there are several Terminals, state-owned and private companies. Consignments of refined petroleum products are bought on the international market by a commodities trader, based on the stated requirements and capacities of the Terminals. The trader arranges shipping and unloading and charges all beneficiaries for their services. In this way, the running costs are shared based on usage and no particular company has the liability of operating and maintaining the facilities. I am sure there are plenty of similar situations across the world.

Another obvious example is the sharing of firefighting services. The Port of Rotterdam has a fire service. All users of the port rely on it and pay for its upkeep. This allows the many Terminals in the Rotterdam area to employ fewer firemen on their site and have the assurance of a well-trained, well-manned and well-equipped fire brigade just a phone call away. The role of the firemen at the site becomes one of the 'first responders' whilst any major fires will be tackled using the manpower and equipment of the port's fire brigade.

One aspect where direct savings can be made is the purchase of foam concentrates. AFFF is expensive and deteriorates over time. It doesn't make sense to store large quantities in each Terminal, but as the fire brigade serves so many Terminals, they can afford to carry large stocks in the immediate vicinity (which can be replenished from stocks in Amsterdam and Antwerp).

The Amsterdam-Rotterdam-Antwerp chemical cluster is fairly unique so despite the obvious benefits, it is not repeated in many other areas (Houston being the obvious exception). Some Terminals share their firewater storage tanks with their neighbours, or agreements are made to provide additional firefighting support to nearby manufacturing plants.

The concept of shared services can extend to any utility on a Terminal where spare capacity exists and neighboring businesses could benefit from access to that resource. Power stations (and Refineries with a power station) may be able to provide low-grade steam for tank cleaning. Local power-sharing agreements can be made. Small Terminals may share marketing and HR resources.

The work of the Maintenance crews tends to be seasonal and has peaks and troughs in workload so smaller specialist Terminals may benefit from having larger neighbours. Whilst some of this work could be done by Maintenance Contractors, the specific skills and knowledge needed in Terminal work are sometimes not readily accessible.

There is one company in Rotterdam that has acquired a liquid bulk terminal, an asphalt terminal, and a break bulk terminal all adjacent to each other. Staff can move around from location to location as peaks in workload fluctuate. HR, marketing, legal, maintenance, and clerical support are all shared between the three locations.

This is the ultimate in utilisation and sharing of resources.

•••••••••••••••••••••

# CHAPTER 3

# *Terminal Engineering*

## 3.1    Heating and Mixing

### Tank Heating

Many products require to be heated before they can be stored (and transported) in liquid form. Traditionally, heating of high viscosity feedstock and residual fuels in refineries was carried out using **steam**.

Most refineries had a power station on site, fired using low-value product or off-gas from the refining process. Unsaturated low-pressure steam was bled from the pass-out turbine driving the generator, and this was used across the site for tank and process line heating. The steam condensed and released its latent heat into the product. As the volume of steam used for this purpose was small compared to the overall volume of demineralised water in the system, condensate from the heating system was not normally recovered. Condensate collecting at the lowest part of a trace heating line was expelled from a local steam trap and discharged to the surface water drain system.

When product temperatures vary below the setpoint, pumping the product can become difficult. When product temperatures vary above a setpoint, the product can be irreversibly damaged. Both outcomes can prove costly.

Tank heating duty requirements vary due to tank volume, insulation level, and initial heat-up time limitations. However, systems are usually designed with a sufficient heating capacity to bring a full tank of product from ambient temperature to its nominal recommended storage temperature.

### *Steam Heating*

There are two basic methods of steam heating tanks:

**Submerged Steam Coils-**The use of steam coils within the tank is particularly common where cargoes of crude oil, edible oils, tallow, and molasses are heated in deep tanks. Many of these liquids are difficult to handle at ambient temperatures due to their viscosity. Steam heated coils are used to raise the temperature of these liquids, lowering their viscosity so that they become easier to pump.

Steam coils are typically continuous lengths of 1" (or similar) low carbon steel pipe, looped across the floor of the tank in a single layer, or installed as a spiral. It is mounted on a framework to keep the steam pipework clear of the floor plates (any debris trapped between the trace pipe and the floor plates can cause hot spots).

Steam enters the heating coil via a control valve. Steam can be regulated to bring the product to its normal storage temperature, which is marginally above its set point. Temperature sensors in the bottom of the tank supply a signal to the steam regulating valve.

Where the product level in the tank is low, the volume of steam is reduced and ultimately shut off. In the same way, there is protection to avoid overheating the product.

Needless to say, tanks that contain heated products are also thermally insulated. It is also not uncommon to have a double tank bottom to avoid heat being leached away from the tank into the ground.

To aid the heating process, mixers are frequently installed on the tank. These can vary between side wall entry mixers (frequently seen on large crude oil storage tanks) and 'bubblers' (used on small asphalt tanks).

The mixers also serve another purpose in refineries. When crude oil is delivered it has quantities of water, sediment, paraffin, and hydrocarbons mixed in, known as 'bottoms'. If this is allowed to settle at the bottom of the tank it will blanket the heating coil, resulting in a reduction of performance and 'coke' forming on the heating coil surface. It can also accelerate corrosion. Therefore, the mixers operate continuously to prevent settling and keep the tank contents homogenous.

Where a greater heat transfer is required, the plain carbon steel pipe can be replaced by a steel pipe with alluminium fins.

**Steam Jackets-**The heating of the product is outside the storage tank in a separate heat exchanger vessel.

The product is pumped from the bottom of the tank to a steam jacket, where the steam is in the vessel and the product is in a pipe passing through the vessel. The heated product is then returned to the tank via a nozzle on the far side of the tank (to prevent recirculation). An alternative arrangement is a shell and tube heat exchanger.

Steam jackets are often preferred these days as they overcome two of the limitation of steam coils. In the event of a steam coil leaking, steam (or more likely condensate) is released into the tank where it either mixes with or displaces the product. With some products, the water can be removed by settling and decanting off the contaminant. In other cases, water in the product may require additional processing to remove it. Either way, a leakage in the steam coil can prove expensive and time-consuming.

The other drawback is that repairs to a heating coil at the bottom of the tank require taking the tank out of service and removing all the product before repairs can be attempted. In tanks carrying crude oil feedstock, black hydrocarbon product, or hazardous waste materials, just getting the tank clean and gas free to enable the start of the repairs can be a long, expensive and laborious task. Steam jackets can be taken out of service and repaired without draining the storage tank.

To guard against the event of a tank's heating being interrupted and the products becoming solid or semi-solid at low temperatures, a seperate heater can be installed inside the tank adjacent to the suction nozzle. When needed these can be used to ensure a quantity of heated low viscosity product is available at the recirculating pump suction.

These suction heaters can be heated by steam, oil, or electricity. Once a small flow of heated product is achieved through the external steam jacket heater, higher flow rates are possible and product temperatures can be gradually raised by continuously recirculating the tank contents. The most common application is heating tanks of asphalt, bitumen, and heavy fuel oil.

## Electric heating

Electric heating has to a large extent superseded steam heating (unless there is a readily available supply of steam already on the site).

Originally conceived in the 1930's, resistance heating cables didn't become available until the 1950s. Self-limiting heating tapes became available some 20 years later.

Electric heating of storage tanks consists of rings of self-adhesive electrical tape attached to the outer surface of the tank shell at regular intervals up to the top. Every so often, a proprietary control unit is attached, which can assign heating limits for either the section of tape or a portion of the tank surface area.

There are different tapes for different purposes. Series element tapes are used on long process lines. Constant wattage tapes have the advantage that they can be cut to length on site. Self-regulating tapes have a resistance that varies with temperature. As the resistance element heats, it allows less current to flow so the cable is inherently power saving.

Whilst electric tank heating has the advantage of being on the outside of the tank (and therefore accessible for repair/replacement) it is under a thermal insulation blanket, vapour barrier, and (frequently) cladding for weather protection. All of this needs scaffolding to remove and replace. So whilst being 'accessible' it is not easy or cheap to repair. Tank heating tapes are staggered so that failure of a single (or multiple) 'loops' is tolerable until the next major outage.

Electric trace heating can be efficient but it still increases electrical demand on the site. As utility prices are predicted to increase over time, tank and pipe heating costs will become larger components of the site's running costs. For larger installations, where a ready supply of low-cost, low-pressure 'waste' heat is available, steam heating may remain a viable option.

## Insulation

Insulation has the following benefits:

- Reduced energy costs for heated tanks;
- Reduced product losses by evaporation;
- Reduced heating means lower GHG emissions;
- Prevents freezing and enables more precise temperature control;
- Insulated tanks are less susceptible to radiant heat in a fire situation;

To avoid corrosion, the thermal insulation material must be water repellent as well as low thermal conductivity. The most popular materials are mineral wool (Rockwool) and fiberglass, which can be supplied in flexible removable blankets, mats, rolls, boards, and pre-formed sections. Mineral wool is denser and more rigid and can be specified to withstand light foot traffic. It is also more resistant to fire, moisture, and mould. Fibreglass insulation can be useful for sealing around penetrations and packing small areas.

Weather protection is usually provided on tanks and pipes by using polythene sheeting covered with aluminum cladding.

Everyone is familiar with insulation on tanks to prevent heat loss but it should also be borne in mind that some tanks have to be insulated to keep heat out. Even in areas where the climate is not so hot, metal in the summer months can heat to 80°C, which leads to evaporation of the contents of tanks and vessels. Economic losses resulting from this are obvious.

Equally important is the problem of insulation of tanks for storing ammonia, sulphuric acid, styrene, and other reactive liquefied gases, which should not be allowed to overheat during storage.

## Pipework Heating

All the pipe work between the tank and the heater must be trace heated as well to avoid plugging of the product in the lines.

Steam heating is usually accomplished by running small bore steam lines along and in direct contact with the piping. The entire assembly is then insulated to direct the heat into the process fluid. As the steam gives up its heat it becomes condensate, which is blown out of the lines at steam traps placed along the pipeline.

Where temperature limits of the heating medium are more critical, **hot oil** was sometimes used. Whilst it reduced corrosion associated with the steam coils and jackets, capital costs of the installation increased. In particular, hot oil heating of pipes was achieved using a 'pipe-in-pipe' configuration which was expensive and time-consuming to fabricate. In short, whereas the steam was the heat source as well as the transfer medium, the oil only acts as an intermediary and must be heated (usually in heat exchangers or fired heaters).

Nowadays, electric trace heating is universally used as it is more efficient.

## Product Mixing

Many products can separate out whilst in storage. Frequently this is the settling of denser components within an unrefined product.

Sometimes products in tanks can stratify with hotter products at the top of a tank and cooler at the bottom. The point of **mixing** is to return the tank contents to a state where it can be considered a homogenous product.

Mixing can be achieved by a variety of mechanical means within the tank or as a consequence of moving the product to a different tank.

The most commonly seen mechanical mixer is the side entry mixer. They consist of an electric motor, drive assembly, and an agitator (like a ship's propeller) on a drive shaft. Most large tanks will have three or four mixers, installed at an angle. When operating, the agitators stir the liquid but also push it tangentially around the tank.

Narrow vertical tanks may have a motor-driven shaft on the centerline of the tank, with rotating paddles (typical for modern asphalt tanks).

On hot or viscous liquids, mechanical agitation is not the best option. In these cases, compressed air is blown into the bottom of the vessel and rises to the top of the tank, causing the recirculation of the product. This method is sometimes used on bitumen blending tanks.

For products that are neither hot nor viscous and do not require aggressive mixing, transfer pumps can be used to recirculate the product within a tank. Alternatively, products can be moved from tank to tank by pumped transfer. To increase the efficiency of mixing, the transfer pump discharge line can have ain-line static mixer installed to impart a 'swirl' to the liquids as they flow through. However, this will have little effect if the tank contents have stratified. In these circumstances, mixing using transfer pumps or recirculation can take a long time to be effective.

## 3.2    Control& Instrumentation

### Cabling

Areas where flammable gas or vapour are present can lead to an increased risk of fire or explosion.

Many liquidsare capable of creating flammable vapours, so it is not uncommon for specific regions within a tank terminal to be classified as 'hazardous'. Sometimes the risk is quite evident, as when flammable liquids are being stored or transported, but in other situations, it is less obvious.

Instrument cabling is used to send control signals to plant and equipment and to send back to the Control Room the signals from the field instrumentation (sometimes in the form of an analogue 4-20mA signal). Because the electricity is not being used to provide motive power, voltages and current are low. However, due to the number of cables, and the potential for accidents and errors, instrument cabling always presents a risk, albeit small.

It is quite common for electrical equipment to have both power and control cables. Electric actuators will have power cables to drive them, but the signals that control the actuator are sent via a multi-core control cable from the MCU (Motor Control Unit).

**Intrinsic Safety** (IS) is a common approach to the design of low power control cables, instrumentation cabling, and field instrumentation going into hazardous areas. The idea is to reduce the available energy to a level where it is too low to cause ignition. That means preventing sparks and keeping temperatures low.High-power circuits such aselectric motorsor non-LED lighting cannot use intrinsic safety methods for protection.

Whilst some types of field instrumentation can communicate wirelessly, the overwhelming majority of field instrumentation is connected with multi-core cables.

To limit the energy going to the field instrumentation, Zener barriers are used. Zener barriers have a diode to limit the voltage and a resistor to limit the current. The barriers are installed outside the hazardous area. The objective is to ensure that under no circumstances will the equipment be able to generate sufficient heat or sparks to ignite flammable vapours. Both the I.S. equipment and the Zener barrier must be certified 'Intrinsically Safe' by BASEEFA, SIRA, or a similar authority. Although the cables for I.S. circuits are not special, it is industry practice for them to have blue-colored polyurethane sheathing for easy recognition and added protection.

As cables and field instrumentation have inductance and capacitance (and hence a capacity to store energy), there is normally a limit on the number of devices on each circuit.

### Field Instrumentation

Field instrumentation is used to measure flow, pressure, and temperature. Without getting lost in the details of how the generic and proprietary technology works, we can summarise as follows:

- **Flow**-flow meters can range from restriction orifices and venturi tubes to more sophisticated technologies (i.e. sophisticated tends to mean more complex, more expensive and proprietary). To be accurate they must have lengths of straight pipe in front and behind the measuring element.

- **Temperature**-the standard practice is to install a thermowell pocket within the pipe work that will accommodate a temperature sensor. These allow the probe to be replaced without dismantling any pipework or shutting down operations.

- **Pressure**-in many cases a simple Bourdon tube pressure gauge provides adequate information for an Operator in the field, but modern process philosophy demands that information should also be available to the process control system. By replacing the pressure gauge with a transmitter that incorporates a local indicator, analogue or digital signals can be sent back to the Control Room.

As measuring levels within the tank is such a large subject, it will be considered in the following section.

## Protocols

The way the field instrumentation signal is transmitted back to the Control Room varies depending on the protocol. Unless you are an Instrument Engineer, it is only necessary to recognise there is a wide variety of protocols (analogue, HART, Fieldbus, Modbus, Profibus plus dozens more).

These are all proprietary and not generally interchangeable, as each supplier attempts to lock users into their specific equipment platform. All equipment suppliers claim to have unique benefits plus operational and cost advantages.

## Measuring Levels

There are many reasons for wanting to know precisely the quantity of product in a tank at any given time. Whilst flows in or out of the tank can be measured by in-line flow meters, modern tank gauging systems are accepted by many Customs authorities as an acceptable alternative. For the required level of accuracy, product temperatures, density, and pressures must also be measured, and the equipment calibrated for each specific tank.

The volume of product in a tank can be useful information for a variety of reasons:

- Product movement – flow rates and spare capacity
- Inventory control – for terminals and refineries
- Custody transfer – when approved by OIML and National institutions
- Overfill prevention – in conjunction with limit switches
- Leak detection - by continuous monitoring of volume

For practicality, we assume the density of the vapour in the headspace is negligible compared to the process liquid. We also assume that the contents of the tank are uniform and homogenous.

### *Manual Dip*

The Operator uses a metal measuring tape to measure the liquid level of the tank through a hatch in the roof of the tank. Originally, manual dips were a legal requirement for custody transfer, and are still occasionally used to verify tank level readings provided by non-contact level transmitters;

### Contact Devices

**Floats** -Only used in small tanks holding environmentally safe or non-flammable liquids. The float is inside the tank and a cable from the float goes to the top of the tank and down the outside via a pulley. There is a vertical scale, with a pointer attached to the cable.

It gives a simple and direct indication of the tank level. Despite its simplicity, it is not unknown for the float to sink (or the cable to become jammed as the liquid level rises). It also cannot be used to give a remote alarm without additional equipment. Because of this, floats are seldom used on modern tanks. More recent float designs have a level transmitter fitted to the roof of the tank to overcome the shortcomings of the old design.

**Gauge Glass** -These can be plain glass or armored glass. Gauge glasses show the liquid level, but if the liquid is uncolored the level can be difficult to read. They also need a backlight to be read at night. Again, these are only used on small tanks with low viscous liquids.

The modern replacement of the gauge glass is the Magnetic Level Gauge. Instead of glass, a metal tube is connected to the bottom of the tank and the headspace. Inside the tube is a float with a series of strong magnets that move a pointer on a scale attached to the tube. This means the gauges can handle high temperatures, high pressures, and corrosive liquids;

**Bubbler**– an open-ended tube extends to the bottom of the inside of a tank and compressed air is forced in. The level of the liquid can be determined by how much air pressure is required to get the air out of the end of the tube.

**Differential Pressure** – a pressure gauge mounted on the outside of a tank on any convenient nozzle (except the pump suction nozzle). The level is determined by subtracting the difference in pressure at the bottom of the tank and in the headspace above the liquid. A visual indicator, calibrated for the specific density of the liquid in the tank, can give Operators a clear indication of the tank's liquid levels. This arrangement has the advantage that the instrumentation is cheap and easy to replace with the tank in service;

### Non-Contact Devices

As technology has improved over the years, the number of ways in which the liquid level within the tank can be measured has increased dramatically. However, the cost has risen significantly as the sophistication of devices has increased.

Refined digital electronics are making level sensors and other measurement devices more user-friendly, more reliable, easier to set up, and less expensive.

Modern instrumentation tends to be of the 'non-contact' type, mounted on the roof of the tank. This has the advantage that the detector head can be changed out with the tank in service and does not become 'fouled' by contact with viscous liquids. A typical method involves measuring the 'time of flight' to sense the distance between the detector and the top surface of the liquid. Different manufacturers prefer different methods or combinations of techniques, but ultrasound, microwaves (radar), and light (laser) have all been successfully used. The choice of methodology is frequently dictated by the liquid in the tank and whether there is any foaming/scum on the top surface of the liquid.

Floating roof tanks may have sensors mounted on the tank shell to measure down to the deck. As the distance between the top of the deck and the liquid level is known, the calculation is straightforward.

## Tank Gauging Systems

All the companies that make level sensors also make Tank Gauging Systems (TGS).

Where a Terminal has only a small number of tanks, getting liquid level information back to a centrally located Operator is straightforward. However, it is now common practice to put the data up on a HMI panel for Operator convenience. If the sensors are from different manufacturers, they may employ different protocols that make the exchange of data more complicated. The situation may become more serious on older Terminals where there is not just a mix of manufacturer's equipment, but also a mix of old and new equipment.

Companies such as Emerson and Enraf sell dedicated Tank Gauging Systems. The objective is for the Terminal Owner to purchase a specific TGS, and after that, all components are 'plug and play'. There are obvious advantages for the manufacturers as it 'locks' users into the manufacturer's product lines when individual products become obsolete and need replacement. All TGS manufacturers claim backward compatibility.

The key benefit of having a TGS is that instead of just the tank level, the Operator sees the mass or the volume of product in the tank. This is achieved using other sensors on the TGS and data about the size and shape of the individual tank. Having all tank data for a terminal centralized means that product movements are simplified and inventories can be reconciled when data from flow meters is also incorporated.

Advertisements for TGS make some impressive claims about inventory control, certified custody transfer accuracy, leak monitoring, and blending. Whether these claims are justified in the long term, only their Customers can judge.

FIG.3.1 In large and modern Terminals, some personnel barely leave the Control Room, using DCS or SCADA screens to operate the plant
PHOTO; Koole

FIG. 3.2 Old analogue instrumentation works perfectly well for smaller Terminals where personnel are outside operating the plant. However, on newer plant or Terminals using automatic control systems, digital instrumentation &transmitters are the only way to go these days

## Limit Switches

Whilst modern tank gauging systems are accurate and reliable, it is good practice to have independent limit switches at both ends of the operating range.

The gauging system will normally indicate when a tank is 100% full, but a separate switch, operated on the physical level of the product in the tank, will be set to give a "High Level" alarm when the tank exceeds 100% of its rated capacity. In addition, there is frequently a "High-High Level" alarm to cater for the occasions a "High Level" alarm has been exceeded but failed to initiate action. Sometimes these will be audio (loud sirens) and visual alarms (flashing lights), but in most cases, the system will be set to automatically trip pumps or close valves to prevent the tank from overflowing.

Likewise, it is good practice to avoid pumps cavitating when emptying the tank. To prevent this separate "Low Level" and "Low-Low Level" switches will be set to initiate alarms and trip the pump before the liquid level drops below the top of the tank's outlet nozzle.

These limit switches tend to be simple, reliable, and robust magnetic level switches, fitted to small nozzles on the side of the tank. They are hard-wired to the control systems and independent of the tank gauging system. Ideally, they should be powered from a non-interruptible power supply for system integrity.

## 3.4    Plant Control Systems

The publicity brochures supplied by the manufacturers suggest that tank gauging systems solve all operational problems. However, it is important to recognize their limitations.

For instance, they are usually straightforward monitoring systems with no ability to control equipment such as pumps, control valves, or EOV. Tank-to-tank transfers, blending operations, and metered discharge cannot be controlled by a TGS. However, TGS systems can transfer data to typical DCS and SCADA systems or terminal automation systems.

For plant automation, there are three main types of systems and a confusing number of hybrids. However, the terminology across the industry is not consistent and manufacturers frequently use trade names rather than industry standard generic terminology. Likewise, Operators can confuse systems and facilities so any description of the system by an Operator should be treated with a degree of caution and clarification obtained by the Engineer or Technician responsible for the system's reliability/accuracy.

A **programmable logic controller** (PLC) has a programmable memory for storing instructions. Common in the 1970s they are less popular now as the 'logic' has to be pre-programmed and downloaded into the PLC memory using a PC. On older versions, changing the logic required the EPROM (erasable programmable read-only memory) to be erased using ultraviolet light. They are still used because they are cheap and reliable and quick to respond but they have limited flexibility.

Where these units still exist, they tend to be used as a controller for a specific piece of equipment, such as a hot oil package or package boiler.

A **distributed control system** (DCS) is a computerized control system for a process or plant.

It allows a Terminal to have a centralized control room where most functions can be monitored and controlled. They are essential where plant automation is anticipated. Operator actions in the field are not required so frequently which allows for reduced manning levels (and reduced operational costs).

Due to preprogramming of critical activities, improved plant safety is claimed and the chances of a product spill are reduced.

It divides the plant into different physical areas and assigns separate controllers to each area of the plant. The whole system is connected to become a single control system with the help of communication buses. Data is acquired, recorded, and logged for report generation and system diagnosis, and process control signals are sent out to the field.

The DCS concept reduces installation costs by localising control functions near the plant area. The advantages of a DCS are a higher level of system redundancy and reliability, but these come with increased capital costs and increased operating costs due to the need for skilled technicians to maintain the system.

DCS systems can be configured to the precise requirements of the Terminal. Each input (from field instrumentation) and output (to a piece of equipment) is assigned a tag number. Operators have a screen with graphics that reflect the plant. The screen can show current data, such as which pumps are operating, which valves are open, and current flow rates.

The Operator can change the status of the plant using a trackball, keyboard, or touch screen.

**Supervisory Control and Data Acquisition** (SCADA) is frequently used for the operation of pipelines. SCADA verifies data to avoid corruption over long distances utilizing two-way communication channels. It is also used for communication between proprietary equipment from different manufacturers. In practice, SCADA sits above the process control hardware and can appear to carry out the same functions as a typical DCS. However, it is a control system architecture and would only control the process by using certain controllers like PLC or DCS. So these controllers would control the field element and the process would be supervised or controlled by the SCADA.

As data handling is more robust with SCADA, the system is used for large-scale processes that can include multiple sites and work over large distances.

## Truck Loading Computers

Where road and rail tankers are loading and unloading, computerized systems can be bought to automate many tasks. Multiple loading bays can be monitored and controlled from a single location, where the Operator has a good view of all loading activities and vehicles. A detailed explanation of product blending at the loading rack is included in **Chapter 6.2**

The controllers can be set to:

- Confirm vehicle ID and available tank capacity;
- Confirm driver ID;
- Check vehicle earthing is satisfactory before allowing loading/unloading to proceed;
- Check vehicle is immobilized, to prevent 'drive away' spills;
- Confirm hose connections are made and vapour recovery hood is in place;
- Check vehicles overfill protection system is linked;
- Start pumps or open control valves;
- Control product flow (ramp up/ramp/down/flush lines);
- Print driver Bills of Lading;

### Terminal Management Systems

There are some Terminal Management Systems (TMS) for use at small stranded depots of a similar design where operational data is uploaded to the internet and merged centrally.

Other than that, the term TMS can mean pretty much what you want it to mean. Usually a software package to run on networked PCs, it does not control plant or equipment. Neither does it accept data from the TGS. Claimed benefits include;

- Product, financial & technical data management;
- Vehicle fleet management;
- Technical monitoring;
- Sales tracking and reporting;
- Sales analysis;
- Business orientated capabilities such as payroll, training records, and recruitment;

As such it can be seen as a business/sales aid to Terminal Operators, although manufacturers such as Siemens may offer more capable software. The situation changes constantly and new capabilities are being developed all the time.

## 3.4    Plant Automation

Ask a Terminal Owner what they want from a new depot and they will say:

- I want every inlet connected to every tank;
- I want every tank connected to every pump;
- I want every pump connected to every tank;
- I want every pump connected to every outlet;

Even on the largest marine Terminal associated with a refinery, you will see a similar thought process:

- I want every product produced by the refinery to be stored in every tank;
- I want every ship unloading position to be connected to every tank;
- I want every tank to be connected to every pump;
- I want every pump to be connected to every tank;
- I want every pump to discharge to every ship loading position;
- I want it all controlled from the Control Room;
- I want to operate 365/7/24 with minimal staffing levels;

Those with an engineering design background will quickly see the flaws in this logic. What the Operator wants is always achievable, but at what cost?

### *Budget/Low Technology Terminals*

I have to say up front that some readers may judge this section to be prejudiced and derogatory to simple or low technology Terminals.

Not at all. I am a firm believer in not 'over engineering' solutions or making things more complicated than they need to be. Elon Musk (he of SpacEx and Tesla fame) has a 5 step design process.

The steps are:

- Step 1: Make your requirements less dumb.
- Step 2: Delete the part or process.
- Step 3: Simplify or optimize the design.
- Step 4: Accelerate cycle time.
- Step 5: Automate

He goes on to explain Step 2 as **'the best part is no part'** (think about it). For a non Engineer, he has some pretty astute ideas about engineering and manufacturing. One of these days they will be running M.Sc. courses based on his thoughts, but back to the topic at hand.

Around the world (in what are generally referred to as 'developing' countries) there are many small Depots in remote locations, with just a handful of tanks. Usually, they handle ground fuels and are used to top up the local filling stations. Sometimes they will serve the agricultural market and have fertilizers or maybe they store chemicals for local industry. The site was probably built at minimum cost. The choice of site and selection of materials/equipment was probably not ideal and driven by availability and cost. They will have low throughputs. They probably only operate in daylight hours. Labour is cheap. They are part of a larger business organization so only have to worry about dispensing products. It is someone else's job (at headquarters) to keep the tanks topped up.

In these situations, the Operator's role is very basic. Switch on a pump. Open a valve. Move a flexible hose. Even though these tasks are basic and repetitive, the Operator must perform their role well to avoid financial and environmental disasters for the Owners. Do not underestimate the people that operate these Depots with minimal backup and support.

When an Operator in a modern Terminal switches a pump 'on', they use computer-controlled equipment and usually only see the icon on a computer screen turn from red to green. What they are missing is the 'real' Operator experience.

When you start a pump and you are standing by the side of it you can;

- Hear it;
- See it;
- Smell it:
- Feel it;

An Operator in a small Depot may not understand much about how a pump is designed but, standing next to it, can tell if there is cavitation, if the coupling is broken or damaged, if the seal is leaking or if the motor is overheating. They have probably worked on the site for years and have started that pump hundreds of times so they know when something out of the ordinary is happening.

It is the same with valves. An Operator in a modern Terminal expects to open and close valves using a switch. But a 'budget/low technology' terminal will probably have just a few valves that require to be operated infrequently. In many cases, the valves will be wedge gates (they are generally the cheapest). Even if a gearbox is fitted, it can take significant physical effort and quite some time to open or close a 12" valve.

It is also possible the 'budget/low technology' Depot is not connected to the electricity grid or has an irregular/erratic supply. Standalone generators are there for essential operations not just as a backup.

As far as flexibility is concerned the 'budget/low technology' Depot does have a distinct advantage.

They are prepared to use **flexible hoses**.

Imagine the following situation. All the outlet pipes from the tanks run towards a pit and terminate at the edge of the pit in flanges or quick-release couplings. The other side of the pit has the inlet pipes to the pumps. In the pit, there are a selection of flexible hoses. Now any tank can be connected to any pump. If the flange or the hose leaks (it will) the leakage goes into the pit.

Now imagine a similar arrangement on the discharge of the pumps, and suddenly you find that any tank can connect to any pump (and any pump can connect to any tank). The Owner's dream has become true-complete flexibility.

For sure it can be messy, it can take effort to move hoses, and it is slow to operate but it is **cheap and it works.**

### Increasing Sophistication

Those who have seen and used flexible hoses in this way know that (with care) it works but the arrangement has several pitfalls. What if the hose is connected to the wrong flange? The wrong tank to the wrong pipe? What if the flange/coupling/hose fails? What if the pit fills (and overfills) with product? What is to stop the entire contents of a tank from emptying into the pit?

Using hoses in a 'snake pit' as described above is probably not a wise decision with flammable or corrosive products. Constant Operator vigilance is essential, and they have to have a way to stop flows quickly in the event of something going wrong. The bigger the pipes, the more difficult this arrangement is.

As Terminals have grown larger and more complex, hoses still have a place but are used mainly for product movement to ships and barges (or between ships and barges) and rail/road tanker loading.

The driving factor in Terminal economics is the cost of manpower. Employers have tried various ways to reduce the cost (sub-contracting and agency staff) but the key is to reduce the number of Operators required to run the Terminal. And to achieve this the Terminals must have various degrees of automation.

Some EOVs are a necessity for firefighting when flammable liquids are stored, but installing EOVs on the inlet or outlet valves of a tank eliminates the need to send staff out into the tank farm. They work quicker than manually operated valves and indicate to the Control Room whether a valve is open or shut. It also minimizes the risk of the incorrect valve being operated (all valves look the same to tired employees at night).

So installing EOVs has distinct operational advantages for Terminal Owners. Manpower levels can be lowered, reducing the Terminal's operating costs.

However, for a new Terminal being constructed, each EOV increased the capital cost of the plant:

- Cost of an actuator;
- Cost of power cabling to the actuator, including trenching and trays;
- Cost of MCU and switchgear in substation;
- Need for adequate power supplies from the grid and maybe larger standby generators;
- Cost of local controls;
- Cost of control cabling between actuator/local controls and MCU;
- Cost of cabling between the MCU and any plant automation system;
- Cost of the increased size of the automation system due to the new actuator;

The situation is similar when installing the facilities to allow an Operator to control pumps remotely. Expenditure on salaries can only be reduced at the expense of an increased capital cost. So ultimately there is a balancing act between the number of Operators and the initial capital cost of the Terminal. That balancing act tends to change over time as labour costs increase.

## The Ultimate

The situation described above assumes only the tank isolation valves and the transfer pumps are automated. With this arrangement, limited pumping operations can be carried out by Operators who are remotely located. However, other circumstances may need to be taken into consideration;

- Changing the lineup of what tank supplies products to what pump;
- Changing the lineup of what pump discharges to what tank;
- Changing lineup for tanks to accept imported products;
- Tank to tank transfers, mixing, and blending;
- Controlling imports and exports by:
  - ❖ Pipeline;
  - ❖ Sea;

The Owner's 'ultimate' Terminal would probably be one where someone phones in all product movements for the week ahead. The plant control system would calculate what products needed to be shifted, and what valves needed to be opened or closed. On schedule, the pumps would start, the transfer would be completed and the pumps would shut down automatically, based on TGS or flow data. Human Operators could be eliminated.

If you are an employer, this is probably the Terminal of your dreams. If you are an employee, this is probably the Terminal from Hell. In practice, such an arrangement may be achievable, but, in reality, would you want it to be?

*Example*

> *We were involved with a new Terminal planned for Rotterdam and the Developers duly issued their wish list.*
>
> *Not surprisingly, they wanted everything connected to everything else. By the end of the Concept stage, the Design Contractors had calculated that they would need a total of 4,000 EOVs to achieve the intended interconnections.*
>
> *Once the implications (practical and economic) had sunk in, a more realistic set of requirements were forthcoming from the Developers.*

Even with increasing automation, human Operators still play a key role in supervising and monitoring plant performance. Humans are amazing multi-tasking machines, who can observe deviations from the norm and come up with creative and practical solutions.

No Owner would dare suggest eliminating Operators because of the potential environmental and safety implications.

## 3.5    Electrical Systems

### Power Distribution

Where possible, Terminals take their primary supply from the National or State grid. This is usually the most robust supply and cheaper than generating the power on-site.

Terminals located in ports will generally take their power from one of the Port transformers. It is not generally considered necessary to get two supplies from two different sources in Europe or the U.S. as National and State supplies are considered reliable and stable. Terminals outside ports will normally connect to the grid at the nearest point with sufficient capacity.

Large Terminals will frequently be supplied from the grid at 11kV, 6.6kV, or 3.3kV. Smaller Terminals will be supplied at 400v (or 415v or 440v as appropriate). The Terminal will have its step-down transformer and switch to disconnect the Terminal distribution system from the grid in the event of an equipment or cable fault on site.

Power supplies on site will be nominally 400 vAC 50 Hz three phase and neutral (electrical supply voltages and frequency vary from country to county).

The feeder is usually laid out as a ring main, around the perimeter of the site. Where electrical equipment is located (such as at a pump slab), a substation will be installed in the vicinity. For security, the electrical systems can be split into two at the switchgear, with separate feeders supplying different sides of the board. The two sides of the board are usually isolated from each other, with a closed 'bus coupler'. If one of the incoming supplies fails, the bus coupler on the busbars opens, allowing the whole system to be supplied from a single feeder. Typically, if there are installed spares (like pumps), Pump 'A' will normally be supplied from the 'A' side of the board, and Pump 'B' will normally be supplied from the 'B' side of the board to balance load.

The substation bus bars provide electrical supplies to individual pieces of electrical machinery. For instance, electric motors will be supplied via an MCU (**Motor Control Unit**). These provide local and remote means of starting a motor, selecting direction and speed, regulating or limiting the torque, and protecting against overloads and faults.

A small motor can be started by connecting it to the power. A direct online (DOL) starter can be used if the current to the motor does not cause an excessive voltage drop in the supply circuit.

Larger motors generally require a motor starter. In the case of 3-phase squirrel-cage motors, they draw a high starting current until it has run up to full speed. This starting current is typically 6-7 times greater than the full load current. To reduce the startup current, motors will have reduced-voltage starters or adjustable-speed drives.

Some motors have a star-delta (series-parallel) starter. These reduce startup current before switching to full power running. 'Soft starters' connect the motor to the power supply through a voltage reduction device and increase the applied voltage gradually or in steps. These soft starters can frequently be part of a variable speed drive (VSD).

Building services, street lighting, etc. are usually supplied at 220/240 vAC 50Hz (one phase of a 3 phase supply) although some countries may have 120 vAC and 60Hz domestic supplies.

In addition, there may also be dedicated low voltage supplies in workshops for hand tools (110v). to increase safety.

High pressure sodium lighting used to be common on Terminals as it was cheap but LED lighting has transformed the market. It is energy efficient and has a long life. Many older luminaires have now been converted to LED, as they can be powered by existing 240v supplies, with the addition of a constant current driver. New LED lighting circuit installations frequently use 24v or 48v.

FIG.3.3 A typical Motor Control Unit panel that will be familiar to many Operators. This one was installed about 1970 and would have had old fashioned relays and control circuits. Newer units are digital with more features.

FIG.3.4 A small pump room for non-flammable products. Larger pumps for use with flammable liquids are usually outdoor and naturally ventilated. When outdoor, the pumps for a particular area are grouped together and referred to as 'pump slabs'.

## Back-Up Supplies

Where a Terminal takes its main supply from the National or State grid, there will frequently be a diesel-powered generator for emergency use. Large generators are permanently installed and hard-wired into the electrical system within the Terminal. Small generators are frequently rental units as this means faulty units can be easily and quickly replaced at the rental company's expense.

Most Terminals use gas oil to power the generator's internal combustion engine. In this case, there will typically be a gas oil storage tank as well as a small gravity feed day tank. It is not common that these vessels will have any installed fire protection. Smaller generators mostly run on standard road diesel (DERV) for convenience, although 'red' diesel may be used if the local taxation rules allow it. When the supply fails, the generator should start automatically and the load transferred to the generator. When supplies are restored, the generator should sense the restored incoming supplies and switch off.

As the generator capacity is much lower than the regular supply, electricity will only be supplied to essential equipment. How 'essential' is interpreted depends on the Terminal's design philosophy.

Some Terminals will only have the firefighting system, the F&G alarms, and critical services such as HVAC classified as 'essential'. In this case the Terminal will stop all fuel handling activities until conventional power is restored.

Some Terminals with pipelines or ship loading facilities will class transfer pumps as 'essential' until the cargo transfer is complete. The bigger the generator, the more liberal the definition of 'essential' becomes.

Where the Terminal is remote or stranded and cannot get a regular or reliable electrical supply from the grid, on-site generators provide the only stable means of operating all the equipment on site. In this case, there would normally be 2 x 100% units.

Plant control systems are computers that operate at low DC voltages and have uninterruptable power supplies with dedicated batteries. Domestic light fittings frequently have battery backups built into the fitting, to allow Operators to continue work or to permit a safe evacuation of personnel from within confined spaces.

An interesting fact is that solar panels and battery-backed lights do not work in Nigeria. When told this by an International Oil Company it took us a while to work out what was meant. What the IOC could not actually tell us was that their Operators steal all the solar panels and backup batteries, to either use at home or sell at the local market. I presume a similar situation exists in other countries, but it is worth considering if designing any plant or equipment for Sub-Saharan Africa.

## Cabling

**Flameproof** and **Explosion proof** are both terms used to mean the same thing i.e. a piece of electrical equipment designed for use in a hazardous area utilizing a heavy-duty enclosure.

ATEX (abbreviation for the French regulation "Equipment for Use in Explosive Atmospheres") applies to explosive atmospheres occurring at atmospheric conditions. In the UK the Chemical Agents Directive (CAD) and DSEAR (the British equivalent of ATEX) cover both elevated temperatures and pressures. ATEX regulations are the ones most frequently applied in Terminals.

Explosive atmospheres are most often the result of an unplanned escape of combustible substances. Once these are released to the atmosphere, they can be deemed to be at ambient temperature and pressure irrespective of their process conditions in confinement.

The cabling of any system employing ATEX or ICEX Exd certified equipment must use suitably rated and mechanically protected cable. The cable must be terminated using Exd-certified cable glands and junction boxes. The glands and junction boxes must be certified Exd by BASEEFA, SIRA, or a similar recognised authority to the same level as the field equipment.

Electrical equipment has to be certified for use in a particular atmosphere:

- **Zone 0** electrical equipment should be intrinsically safe (EN 50020) or Ex S. As a consequence, electrical equipment is seldom installed within Zone 0;

- **Zone 1** electrical equipment is
  ❖ 'd' (flameproof EN 50018);
  ❖ 'p' (pressurized EN 50016);
  ❖ 'o' (oil immersed EN50015);
  ❖ 'ib' (Intrinsic safety EN 50020) or
  ❖ 'm' (encapsulation EN 50028);

- **Zone 2** electrical equipment is type 'n' EN 50021;

In addition, electrical equipment has to be specified T1-T6 depending on the ignition temperature of the gas or vapour it may be exposed to in service.

When the hazardous areas of a plant have been classified, the remainder will be defined as non-hazardous, sometimes referred to as 'safe areas'. The zone definitions take no account of the consequences of an unplanned release.

The UK Conformity Assessment ("UKCA") marking applies to most products previously subject to the European Union's CE marking. UKCA marking came into effect on 1 January 2021. However, many companies have dispensations to continue with their CE marking until 1 January 2023.

The UKCA mark may not be on the item, but may be attached to packaging or the accompanying documents e.g. manuals.

# CHAPTER 4

# *Loss Prevention*

## 4.1   Tank Integrity

### Unplanned Releases

Tanks generally have long, trouble-free service lives. Unsurprisingly, when tanks fail they are well documented.

The United States Environmental Protection Agency (USEPA) commissioned a study to investigate the common sources of an **unplanned release**. The study covered ten years (1990 - 2000). Of the 312 incidents during this period it was found that the root causes of the releases were:

- 55% tank failure;
- 22% operator error;
- 10% to valve failure;
- 4% to pump failure and
- 3% to bolted fitting failure;

### Causes of Tank Failure

As tank failure is the major reason behind these unplanned releases, it is worth reflecting on the factors that cause a tank to fail in service:

- Poor Design;
    - ❖ Incompatibility of fluid with design;
    - ❖ Seismic effect;
    - ❖ Subsidence of foundation;
- Poor Construction;
    - ❖ Welding issues;
    - ❖ Foundation construction;
    - ❖ The inexperience of the Contractor;
- Poor Maintenance;
    - ❖ Corrosion;
    - ❖ Defective instrumentation;
    - ❖ Defective PRV (Pressure Relief Valve);
    - ❖ Defective VRV (Vacuum Relief Valve);
    - ❖ Joint/gland packing failure;
- Poor Operation;
    - ❖ Incompatibility of fluid with tank materials;
    - ❖ Ignition of flammable gases (explosion);

- ❖ Dispensing;
- ❖ Overfilling;
- ❖ Vacuum (implosion);
- External Events;
  - ❖ Sabotage;
  - ❖ Exceptional weather;
  - ❖ Lightning/static electricity;
  - ❖ Fire of adjacent tank;
  - ❖ Mechanical damage (impact);

## Types of Tank Failure

- Shell/Roof (frangible joint) failure;
- Shell/Bottom joint failure;
- Bottom failure;
- Shell plate weld failure;
- Structural failure due to heat damage from a fire;
- Rapid failure (unzipping);
- Foundation failure (leading to shell/bottom joint failure);
- Nozzles/connecting pipe work;

## Typical Tank Defects

We frequently see carbon steel tanks up to 80 years old still in service, so it is no surprise the most frequent defect is corrosion. Carbon steel tanks in hydrocarbon service typically have corrosion in the following areas;

- Conical tank roofs tend to buckle at the joint with the wind girder/shell plates. As a result, rainwater can 'pool' around the perimeter of the roof. Over time the constant wetting and drying of the roof cause pinhole corrosion on the exterior of the roof plates, leading to perforation. This process is accelerated when the tanks are in a marine or coastal environment.

- Even in hydrocarbon service, water accumulates inside the tank. As the tank is emptied and filled, and due to the changes in temperature between night and day, the tank 'breaths' and condensation can form on the inside. Rainwater can enter the tank. Heating coils in the tank can leak, allowing steam into the tank, which can then condense. Unprocessed hydrocarbon cargos can also have a quantity of BSW (bottoms, sludge, and water).

  The consequence is that two specific areas of the tank's interior can suffer accelerated corrosion. EFR tanks tend to have a band of corrosion all around the shell about 1–2 meters above the floor plates. Also, due to bacterial (anaerobic) corrosion, the interior surface of the floor plates can show a pronounced metal loss.

- The annular ring and the floor plates generally sit on a foundation (see below) and it is not uncommon for the water to penetrate by capillary action, causing corrosion on the exterior of the floor plates.

Most corrosion can only be found with detailed inspection whilst the tank is out of service.

## Tank Inspection

For want of stating the obvious, tanks (and many other plant items) are inspected to avoid an unplanned failure. Any failure could release large quantities of liquid that could be combustible, flammable, corrosive, or toxic, lead to a vapour release and explosion, or cause a major incident in some other way.

Most inspection regimes require an **annual** external inspection of the tank and foundation by a suitably qualified Tank Inspector.

Internal inspections with the tank out of service are usually required on a fixed schedule (every 10 years typically) or as calculated if the Owner has adopted a **Risk Based Inspection** (RBI) philosophy. RBI is looked at in more detail in **Chapter 4.4-Risk Based Inspection.**

A full list of inspection items is contained in EEMUA 159 but the significant items have been included in **Appendix D** for easy reference.

While the inspection requirements do not seem too onerous, it must be borne in mind that before any of these actions can take place, the following has to be completed:

- Tank stripped of the product;
- Vented and made gas-free;
- Man access doors opened;
- Emptied of residue and sludge;
- Tank cleaned internally;
- Scaffold erected internally;

## Tank Inspection Standards

- EEMUA 159          Risk-based tank inspection;
- API 575          Inspection of Atmospheric and Low-Pressure Tanks
- API 580          Risk-based inspection;
- API 653          Tank inspection/repair;
- API 2350          Overfill prevention;
- API 2000          Venting;
- STI SP001          Standard for Inspection of Above ground Storage Tanks

## Tank Inspection Methods

Out-of-Service inspections of tanks to EEMUA 159 and API 653 can include using the following techniques:

- Ultrasonic Thickness survey;
- Eddy Current Inspection;
- Magnetic Particle Inspection;
- Dye Penetrant Inspection;
- Vacuum Box Testing-for floor plates;

## 4.2   Pipework Integrity

### Pipework Inspection

API 570 was issued to cover the inspection, rating, repair, and alteration procedures for metallic piping circuits that have been in service. Whilst many larger Terminals claim compliance, small Terminals tend to adopt a less formal approach.

A typical Integrity Assessments to API 570 report may include:

- An overview of the condition of the piping circuit including ;
    - ❖   the recommended inspection interval and
    - ❖   next inspection dates.
- Recommended actions to bring the piping system into compliance with the inspection codes;
- Indication of remaining life based on pipe design and inspection codes;

Where inspection of pipework takes place, it tends to be on an area-by-area basis, or system-by-system basis. It is therefore difficult to make any general statement covering the condition of all the pipework on a typical Terminal.

To the casual observer walking around the site, the condition of pipework is frequently assumed based on its external appearance. Whilst understandable, this is misleading as it assumes the majority of the corrosion and metal loss is on the exterior of the pipe. One of the things the 'casual observer' needs to recognize is that in any Terminal there are bound to be some sections of pipework no longer in service, usually because they are **redundant**. Terminals do not tend to demolish redundant pipework as you never know when you may need it again in the future. However, there is no point wasting money maintaining lines that are not in service, so they are not repainted with the same frequency (or at all). Where pipes run along a pipe track, weeds and grass take over, adding to the air of dereliction and keeping the pipe wet (adding to the external corrosion).

### Pipework Inspection Standards

Inspection must be carried out by certified inspectors, trained in each specific technique (i.e. one qualification doesn't cover everything). Typical accreditations are:

- Certification for Welding Inspection Personnel (CSWIP) UKAS accredited to ISO/EC 17024
- Personnel Certification in Non-Destructive Testing (PCN) meets BS EN ISO 9712
- American Society for Nondestructive Testing (ASNT) covers 5 NDT test methods

### Pipework Inspection Methods

There are many NDT techniques suitable for the inspection of pipework, including:

- Ultrasonic Thickness survey-detecting sub-surface flaws:
- Eddy Current Inspection-detecting sub-surface flaws:
- Magnetic Particle Inspection-Uses magnetic fields to find discontinuities at or near the surface of ferromagnetic materials:
- Magnetic Flux Leakage-a sensor is used to detect changes in magnetic flux density to show any reduction in material thickness:

- Dye Penetrant Inspection-mostly used to locate surface defects:
- EMAT (Electro Magnetic Acoustic Transducer). For quantifying corrosion at pipe support touch points:
- Long Range Ultrasonic/Guided Wave:
- Radiographic Inspection-X-rays are commonly used for thin or less dense materials while gamma rays are used for thicker or denser items

Using these techniques, Inspectors &NDT technicians can identify metal loss (from corrosion or erosion) on either the inside or the outside of the pipework, even when the pipe is below ground level.

If a line is suspected of leaking, the following tests can be used:

- Bubble leak test- uses a tank of liquid, or a soap solution for larger parts, to detect gas (usually air) leaking from the test piece
- Pressure Change Testing-a loss of pressure or vacuum over a set period will show that there is a leak
- Halogen Diode Testing-vessel or pipe pressurised with tracer gas and detected with a gas 'sniffer'

## 4.3    Pressure Vessel Inspection

In the UK, the **Pressure Systems Safety Regulations** 2000 (PSSR) cover the safe design and use of pressure systems. PSSR aims to prevent serious injury from the hazard of stored energy (pressure) as a result of the failure of a pressure system or one of its parts.

The PSSR requires pressure systems to be inspected in accordance with a Written Scheme of Examination.

In Europe, three directives form the legal framework for pressure vessels:

- **Pressure Equipment Directive** (PED 2014/68/EU);
- **Transportable Pressure Equipment Directive** (TPED 2010/35/EU) and
- **Simple Pressure Vessels Directive** (SPVD 2009/105/EC).

Although most pressure vessels in the USA have been built to Section VIII of the ASME Boiler and Pressure Vessel Code (BPVC), the actual requirements for the design, manufacture, and operation of pressure vessels are defined at the state level. This means that not all states have the same requirements. Most states require that pressure vessels are inspected to either the API 150 Pressure Vessel inspection code or the National Board Inspection Code.

The regulations apply to owners and users of pressure systems containing steam, gases under pressure, and any fluid kept under pressure that becomes a gas when released to the atmosphere.

## 4.4    Risk-Based Inspection

In terminals, tanks and pipework are the primary containment for the stored liquids. The objective of any inspection system is to ensure it retains its integrity between out-of-service inspections.

The frequency of inspections of tanks and pipework was once determined by prescriptive codes. The inspection frequency, method of inspection, and areas to be examined were determined with little consideration to the plant's age, duty or condition.

Increased operational experience has led most of the industry to adopt a more reasoned approach to inspection planning. The objective is to inspect the tanks and pipe work at a frequency appropriate to their actual condition and duty.

**Risk-based inspection** is a philosophy that considers the likelihood (probability) of failure due to flaws, damage, deterioration, or degradation with an assessment of the consequences of such failure.

When applied to primary containment, the information is used to identify;

- The type of damage that may potentially be present;
- Where such damage could occur;
- The rate at which such damage might evolve, and
- Where failure would give rise to danger;

A suitable inspection scheme will determine the optimum frequency at which inspections should take place.

The principal document used in establishing an RBI inspection regime for storage tanks is **EEMUA 159 "Users' Guide to the Inspection, Maintenance and Repair of ASTs"**. The Engineering Equipment and Materials Users Association (EEMUA) is a UK-based trade body but this particular specification has become the de facto standard for all tank inspection systems.

In addition, the following also give guidance on how to implement an RBI inspection philosophy on all manner of plants and equipment;

- API 579        Fitness for Service
- API 580        Risk-Based Inspection
- API 581        Risk-Based Inspection Technology

It is important to recognise the limits of these standards. They give guidance on setting up the inspection system. They do **NOT** lay down the standards for inspection techniques and competency requirements. Typical of the standards that deal with these aspects are:

- API 510        Pressure Vessel Inspections;
- API 570        Piping Inspections;
- API 653        Tank Inspections;
- API 2610        Design ......... and Inspection of Terminal and Tank Facilities

The terminal industry sees RBI as a way to extend the time between (costly) out-of-service inspections and minimise the inspection, cleaning, and scaffolding costs. Regulatory pressures ensure that RBI is carried out rigorously so that inspection decisions are based on adequate information and expertise.

## 4.5    Fire Protection

Fire protection only needs to be considered at terminals storing flammable products.

Even then, there are probably many hundreds of small ground fuel terminals that do not have any installed fire protection. Many of these terminals will be in remote locations, without staff trained in fire fighting and no ready supply of firewater. In these cases, in the rare event of a major fire, the tank and its contents will be written off, and hopefully, damage to the rest of the equipment will be minimal.

Tank farm fires in the oil and petrochemical industry do not occur often (most of the headlines are created by the process plants where temperatures and pressures are much higher). When they do occur, it is with devastating consequences and negative publicity.

## Fundamentals

For a fire to occur needs three things:

- Oxygen;
- Heat;
- Fuel;

Typical sources of ignition are;

- Flames;
- Direct-fired space and process heating;
- Cigarettes/matches etc.;
- Cutting and welding flames;
- Hot surfaces;
- Mechanical machinery;
- Electrical equipment and lights;
- Sparks;
- Stray currents from electrical equipment;
- Electrostatic discharge sparks:
- Lightning strikes;

The following codes/standards are frequently referred to during the design of a terminal:

- NFPA 30          Flammable and Combustible Liquids Code;
- NFPA 11          Low, Medium, and High Expansion Foam;
- API RP 2001      Fire Protection in Refineries;
- API Pub. 2030    Water Spray Systems .... in the Petroleum Industry;
- API Pub. 2218    Fireproofing Practices in Petroleum and Petrochemical ....;
- NFPA 16          Deluge Foam;

## Tank Spacing

The requirements for tank spacing and fire protection may be covered by national codes or company standards. These will have been agreed with the terminal's insurers so should be a reliable guide to the minimum acceptable standards. However, many of these standards are based on **NFPA 30 "Flammable and Combustible Liquids Code"**.

This code gives nominal spacing between vertical storage tanks holding different classes of product.

For Class 1 products, a distance of half a tank diameter is recommended as the space between adjacent tanks. This distance is to protect adjacent tanks from the radiant heat of a tank on fire. Also specified is the volume of water discharged over the tank roof/shell of the adjacent tank for cooling in the event of a fire. The standard also specifies the minimum distance from a tank to the property boundary line.

There is purpose-made software for optimising these tank spacings. Shell Oil Company originally developed a software package called "FRED" (fire, release, explosion and dispersion) for internal use.

This modeling tool is now available as a commercial software package from Gexcon, and the results generated could be used to justify different (reduced) tank spacing.

## Fire and Gas Detectors

The best fire and gas detectors on any Terminal are the Operators who know the site and the specific risks. Traditionally, hard-wired 'manual call points' (break glass units) are located at strategic points around the site, but as most Operators carry a radio when 'out and about', this has now become the default method of raising an alarm.

Smoke detectors are uncommon in industrial plants as the major risks are out of doors. Smoke detectors are typically used inside low risk manned buildings, sometimes in conjunction with 'rate of rise' detectors. Manned buildings will also commonly have monoxide gas detectors.

The major risk with open top tanks is a rim fire caused by lightning or static electricity. The use of linear heat detection cables between the shell and the foam dam is commonplace these days. The old method of compressed air in plastic tubing (the tube melts when it gets hot, and the fall in air pressure triggers the foam generators and alarms) is still around.

Flame detectors can be used to identify the UV or infrared signature from a fire along a line of site. Optical detectors are also available. Again, there are combined detectors to identify flames, heat and smoke.

Gas detectors are fitted where there is a risk of flammable and toxic gases and vapours being present. In most cases the Operators will carry this $CO_2$, $H_2S$ or LEL detectors (depending on the products stored and the identified risk) when they go out into the Terminal.

## Fuel Transfer from a Tank on Fire

Assuming a fire has started in a tank, the best way of minimising the duration of the fire is to transfer the contents of the tank to another tank outside the risk area.

The value of the product in the tank may exceed the capital value of the tank itself. Also, during any firefighting effort, water and foam may contaminate the product in the tank. Therefore, it makes economic sense as well as practical sense to save as much product as possible.

There is one caveat to this. Large external floating roof tanks are used to store crude oil. A boilover can occur in crude oil fires when the 'hot zone' of dense hot fuel created by the fire reaches the water at the bottom of the tank. The water turns to steam, which pushes through the crude, taking fuel with it and creating a fireball above the tank. Boilovers spread the burning crude, thus escalating the incident and endangering the fire responders. In many instances, the eruption of burning crude oil has been violent enough to create a domino effect and set adjacent tanks on fire.

Transferring the crude to another tank is still the preferred option, but Operators should be aware this may make the boilover happen sooner rather than later.

The current approach, to make this tank-to-tank transfer possible, is to fit an FSV (fire safety valve) on an outlet nozzle of the tank. The FSV is usually a quarter-turn valve with valve seats designed to operate under fire conditions. The valve has either an air/spring actuator or an electric actuator. To protect the actuator, it frequently has a fire-resisting blanket and any power/control cables have radiant heat protection.

In the event of a fire, operators can remotely open the FSV and use the transfer pumps to salvage the contents of the tank before they are contaminated or destroyed.

## 4.6    Fighting a Tank Fire

How operators attempt to put out a tank fire depends on many variables.

### Fixed Roof Tanks

If the tank has a fixed roof, the fire must be tackled by spraying the extinguishing foam inside the tank. Depending on the specific gravity of the liquid in the tank, the foam connection is either at the top or the bottom of the tank. When there is a top foam connection, to prevent vapour from escaping through this connection there is a glass panel, which breaks under pressure allowing the foam to flow in.

As the extinguishing foam is less dense than the product, it will float on top of the liquid. This will therefore create a shield against oxygen and the fire will be extinguished.

If there has been an internal explosion in the tank, the chances are the roof will have become partially or fully detached at the 'frangible' joint between the roof and the shell. Injecting foam may still work but now it may also be possible to use high-pressure monitorsto send a jet of aspirated foam into the space created by the roof joint failureand onto the burning liquid.

There is no simple way to put out a fire in a storage tank. The longer the fire goes on, the more established it becomes and the more difficult it is to put out. Shell and roof become so hot they can re-ignite the fire. Also (obviously) until the advent of remote drones, it was not possible to see over the lip of the tank shell from ground level, so fire-fighters were effectively working blind. Although some fire brigades will now have drones available it is unlikely Terminals will have access to similar equipment.

The objective of the firemen will be to get enough foam into the tank to overwhelm the fire. Initially, they will start by directing the stream of foam over the top of the tank and against the far wall of the tank. Assuming the heat from the fire doesn't break down the foam before it can do anything, the foam will spread across the surface of the product. The size of the pool fire will gradually diminish until the fire is quenched.

Whilst this all sounds straight forward there are some complications. The firemen need enough pressure in the hydrant system to direct the stream of foam across the tank but they also need a sufficient flow rate. In many cases, especially with large tanks, the firemen will have insufficient pressure to get the water over the lip of the shell and across a burning pool of liquid.

A typical hose is about 50mm in diameter so special foam projectors have been designed, using multiple hoses to supply the volume of water required. However, these devices are relatively uncommon. In addition to supplying water to fight the fire, the hydrant system may be supplying cooling water to adjacent tanks. Another problem is that foam is a limited resource at most terminals. Foam concentrate is not cheap and deteriorates with time, so terminals tend to keep limited supplies and rely on 'group stocks' held elsewhere if needed.

## Floating Roof Tanks

Fires in floating-roof tanks are usually in the area between the tank shell and the floating roof, and a frequent cause is a lightning strike

A modern solution is a rim seal extinguishing system combined with a linear heat detector cable.In the old days, the detection was done by plastic tube holding compressed air. When it got hot enough, the tube melted releasing the air and starting the foam system.

The rim fire fighting system is designed to smother the fire by dumping aspirated foam into the space between the tank shell and the foam dam. The foam dam, consisting of a short vertical plate, is welded to the top pontoon plate at a short distance from the seal. The height of the foam dam is higher than the upper tip of the seal; thus allowing the whole seal area to be flooded with the foam and extinguish the fire effectively.

The foam equipment package can be mounted on the floating roof, or the foam can be generated by a package at ground level. Risers take the foam to the foam pourers at the top of the tank.**NFPA-11 for Low-Medium-High Expansion Foam** has a detailed explanation of how to design such a system.

In a fire situation, the 'worse case' scenario is the floating deck sinking or capsizing. In this event, the firefighters are reduced to using the tactics used on fixed roof tanks, but with little chance of extinguishing the fire.

The lessons learnt by the LastFire Steering Group include:

- A boilover event will happen in every case where there is a full surface fire in a crude oil tank;
- Ejected boilover product is usually contained within the secondary containment, although can spread up to 10 tank diameters downwind, spreading the fire to adjacent tanks;
- Boilover can still occur even when the fire has been extinguished and may happen more than once;
- Three elements must be present for a boilover to occur:
  - ❖ A full surface fire;
  - ❖ A water layer or pockets within the tank;
  - ❖ The development of a 'hot zone' (only happens with crude oil);

FIG. 4.1 Amoco Milford Haven crude oil fire after boilover had spread the fire to adjacent tanks
PHOTO; Wetern Daily Mail

Fig. 4.2 A chemical tank fire in Jurong. The sides of the tank crumpling is common where the liquid level has been reduced.
PHOTO; Jurong Aromatics Corp.

The ideal window of opportunity for a concerted foam attack on the fire is a matter of a few hours. Ideally, foam should be initiated within 2-4 hours, but in practice this is seldom achievable and a 'burn down' policy should be adopted. The only viable strategy is to set up cooling of adjacent structures, pump out crude if possible and withdraw firefighters to a safe distance to await the boilover.

Although seldom acknowledged, it is commonly accepted within the firefighting community that extinguishing a major, fully developed fire on a large crude oil tank is extremely unlikely. Any attempts to fight the fire may be entirely for public relations purposes.

## 4.7 Fire Related Topics

### Hydrant Systems

As noted above, the ability to fight a tank fire is dependent on a hydrant (ring main) system with dedicated fire pumps. The hydrant system is normally charged with water and kept pressurised at about 3 or 4 bar. When water is taken from the ring main, the system pressure drops and the fire pumps start operating in sequence until a system pressure of 11 or 12 bar is achieved. The fire water system design usually ensures a pressure of 7 barg at the farthest point of the hydrant system, however, this may vary depending on the design codes used.

If a secure supply of electricity is available, electrically driven pumps are to be preferred as they have higher availability and reliability. Alternatively, the pumps can be powered by direct drive diesel engines, but this requires a diesel day tank (which may also need installed fire protection).

If the system is charged with sea or river water the internal corrosion on steel lines can be dramatic and frequent flushing is advised. In this case, it is better to have the hydrant system made from GRP to avoid corrosion.

To avoid the use of seawater in the hydrant system, some terminals have a fire water tank, topped up from potable water supplies. A 'typical' fire water tank will have a capacity sufficient to run the firefighting system under fire conditions for 4 hours, before resorting to a river or seawater source of supply. It is good engineering practice to have two 50% capacity tanks and some design codes specify this.

The fire water storage capacity and the fire pump design throughput will be calculated on the assumption that the Terminal, assisted as necessary by outside agencies, will have to fight (or at least contain) a fire on the single largest risk on the site. Having multiple fires simultaneously is not considered a feasible design scenario.

The hydrant system will have the fire pumps located well away from the risk area. The hydrant main runs around the perimeter of a tank pit, usually underground. Where a number of tank pits are being protected the hydrant will run around the outside of the area as well as around each tank pit. A well-designed system will have isolating valves that allow small sections of the line to be taken out of service without losing all fire water supplies to the remainder of the tank farm.

Water for tank cooling and firefighting purposes is taken from the hydrant system. Along the route of the hydrant, hose connection points (with standard fire brigade hose couplings) are installed. Hose connection types and sizes vary between countries so some understanding of local standards is helpful.

In the event of a fire, trained Operators or the local Fire Brigade can connect their hoses and equipment to the site's permanent fire water system. Less frequently spaced will be the permanent fire monitors, allowing water jets to be directed at the tanks in the pits. The spacing of the monitors will be decided at the design stage, based on the tank spacing, layout, and monitor 'throw'.

NFPA 11 and NFPA 16 give guidance on monitor coverage but monitors between 30m and 45m apart is a typical requirement.

## Radiant Heat Protection

An objective of any fire protection system will be to stop other tanks adjacent to the fire, from igniting. The primary cause of these sequential fires is the radiant heat produced during a fire. To minimize the risk, adjacent tanks are cooled with water. NFPA 30 stipulates the quantity of water per square metre but does not stipulate how the water is to be applied, so various methods are applied.

Some tanks have spray rings with spray nozzles around the outside of the tank. In some cases there are multiple spray rings, at various heights. Sometimes there is a single ring around the perimeter of the tank or with some cone tanks, at the centre of the roof. Water cascades off the roof to cool the tank shell.

Where economy with water is a priority, the spray rings can be designed to protect only the quadrant or quadrants closest to the adjacent tank, however, this does cost more to install and adds complexity.

Strictly speaking, as the cooling water sprays should never be used in anger, there is no obvious reason why brackish water couldn't be used instead of the (fresh) water in the fire hydrant system. In practice, this hardly ever happens. The corrosion of pipes carrying seawater is ferocious, meaning that you can really only use GRP piping.

In addition, Operators don't like the idea of their tanks getting dirty which is inevitable if you spray them with dirty, rust coloured water that has been in the pipe for the last year (or more!).

## Fire Monitors

Fire monitors are designed to deliver large water flows in high-risk or hazardous situations. They come in a variety of sizes and of varying complexity.

Smaller monitors can be manually operated and have swivel joints to allow the jet of water or foam to be directed vertically and horizontally. Water oscillating monitors move side to side using water pressure as the energy source. Monitors designed for high-risk areas can be electrically operated, usually by remote staff using CCTV for guidance.

The advantage of monitors is their flexibility. They can be directed as required so can be used for tank cooling, tank fire fighting, and fire fighting in the secondary containment. They can be specified with adjustable nozzles, allowing the operator to select a jet of water or a fog to provide radiant heat protection. It is also possible to specify the monitor with an air aspirating nozzle for use with foam concentrate. The flow of water through the monitor self induces the correct flow of foam concentrate to produce the expanded foam at the nozzle outlet.

Brochures for fire water monitors always include details on the height and throw of a jet of water. It is only in practice you see that both of these are significantly affected by the wind. It is not unknown for a jet of water to break up midflight as if it had hit an invisible wall. As with most things, practical experience for the Operators is vital to understand the physical limitations of the equipment.

## Fire Fighting Foams

Foam is considered the best extinguishing agent for large liquid fires.

Foam can exclude oxygen from the fuel vapours, separate the flames from the fuel, cool the fuel surface and prevent the release of vapours from the fuel. Firefighting foam is created using a foam concentrate mixed with water and aerated.

Fighting fires in tanks and bunds requires Class B foams that can be protein-based or synthetic. The current standard in hydrocarbon terminals is the use of aqueous film-forming foam (AFFF) which is a low expansion foam. Foam concentrate comes in different qualities. A 1% foam concentration will be mixed 1 part of foam to 99 parts of water and expand up to 20% after suitable aeration.

There are several ways of adding foam concentrate to the hydrant water. Some hydrant systems have a centralised foam package that injects the foam concentrate into the water on initiation of the system. Typically, these will have bladder storage tanks (allowing water to pressurise the foam supply). There will be a proportioner to make sure the correct quantity of foam is added to the water. The foam/watersolution then flows through the hydrant system and is aspirated as it leaves the monitors.

An alternative is to have a water/foam premix already in the hydrant system, Historically, these pre-mixes deteriorated over time, but foam chemistry improves over time and this may no longer be an issue with certain brands.

Foam concentrate can also be stored in a bowser or tanker, and when required can be towed to where it is needed. An alternative is to have a small quantity of foam positioned at each fire monitor for ready use. Typically, the foam can be supplied in 600 or 1000 litre IBC (intermediate bulk containers) on steel or wooden pallets. When the foam IBC is exhausted, it can be swapped for a fresh IBC from elsewhere on the site. Storing foam concentrate in small, air-tight containers is the best way to extend its life.

Use of foam, even in a fire situation, is not without environmental dangers. During the Buncefield fire, large quantities of foam found their way down to the water table and there was significant concern about potential long-term consequences.

The use of High Expansion foams to fight minor bund fires caused by a product spill is considered in **'Spills/Bund Fires'** below.

## Extinguishers

The use of extinguishers is mostly limited on Terminals to manned buildings. Portable handheld Dry Powder extinguishers are suitable for Class A, B and C fires. Portable handheld $CO_2$ extinguishers are suitable for Class C fires.

Class A fires include trash, wood and paper. Class B fires include liquids and gases. Class C fires include energized electrical sources.

It is not uncommon to find large, trollied Dry Powder extinguishers at vehicle loading bays, where Operators and Drivers can use it to 'knock down' a fire before it develops. An alternative is a 4 wheel trailer with vessel containing dry powder and a separate pressurized AFFF vessel for the larger fires. These are usually equipped with a very long hose on a hose reel so the Operator does not have to move the trailer closer to the source of a fire.

## Electrical Buildings

Theoretically, the electrical installations on Terminals should be a major source of fires as they can generate the heat necessary. In practice, electrical fires at Terminals are relatively uncommon.

Industrial switchgear is robust and designed to high standards-even the old stuff! Most electrical motors have over-temperature sensors fitted to the windings, which means the motors are electrically isolated by the motor protection circuits before dangerous temperatures are reached. The switchgear (or MCU) will also protect against short circuits on HV and LV systems.

Individual cables are earthed as they enter the switchgear (using approved glands on the gland plate) so cable faults can be identified and the power isolated by tripping the affected circuits.

With so much electrical equipment in one location, the sub-stations would appear to be a potential risk area. Historically, these buildings would be protected by an inert gas or Halon. However, the use of Halon is now banned as it was found to be carcinogenic. Where sub-stations have installed fire protection these days it is invariably $CO^2$. It has the advantage that it leaves no residue and does not damage electrical equipment. The downside of $CO^2$ is that it is an asphyxiate and potentially fatal in confined spaces. Because of this, it is standard practice to lock off the extinguisher bottles before entry to the building. This is not such an onerous element as most substations require a PTW to enter.

## Spills/Bund fires

Often planned for but seldom happening, there is a scenario of product spilling into the secondary containment and catching fire. However, this assumes two coincident events-a spill of flammable liquid and subsequent fire.

There are two distinct stages-contain the fire and then extinguish it.

Many tank pits containing volatile liquids have dwarf walls around each tank within the secondary containment. A dwarf wall can be either a concrete wall, a dyke wall or a modular containment system. The objective is to contain minor spills, not to hold the full contents of the tank. In the event of a leaking flange or valve, the dwarf walls are there to stop burning liquid from moving around inside the secondary containment and spreading the fire to other tanks.

Dwarf walls are not without problems. Concrete dwarf walls are usually inverted tees and installing these during construction can be a problem due to the disruption within a congested area. They also restrict the movement of heavy machinery within the bund, both during construction and during routine maintenance. The easy solution is to form an earth ramp either side of the dwarf wall.

As with tank fires, AFFF is the best option for putting out small bund fires. The best way of fighting the fire is by the rapid reaction of first responders, using the fire hydrant system and the monitors spaced around the tank pit.

Some locations have adopted a different approach and installed high expansion foam generators along the edge of the secondary containment. In the event of a fire, foam is generated and pushed into the bund. The logic is that the foam will smother any small fires by stopping oxygen getting to the product.

I am yet to be convinced of the effectiveness of this approach. The only force pushing the foam along the ground is gravity. In a confined space (such as a cable tunnel), fans can 'push' a wall of foam along but there are limits to how far it will travel. In addition, if there is a fire in the tunnel the heat evaporates the water in the foam and this further constrains the foam's ability to overwhelm a fire.

Using foam in an open bund means that it is at the mercy of the winds and weather. On some occasions when I have witnessed tests of foam discharges into a bund, more foam has ended up outside the bund than in the bund.

## Water Sprays/Sprinklers

Some risk areas in larger terminals may have water sprays.

Sprinklers are designed for low-pressure systems. They are frequently associated with the cooling water sprays on the tanks but they can also be used to protect small horizontal vessels and tanks. Each sprinkler is designed to emit a cone of water and this is suitable for wetting large areas. They are positioned above the item deemed to be at risk from radiant heat, protection of the underside being from water flowing over the structure. Actuation is usually achieved using a linear heat detection cable or a fuse link. A fuse link is a connector made of a material that melts at a prescribed temperature. When the fuse link melts, a spring-loaded valve is released and the water deluge starts.

Water sprays are designed for use at higher water pressure and give a jet at the nozzle. They are directional, so when placed over cable trays the water can be directed along the tray for a considerable distance.

Some Terminals use sprinklers and sprays to protect the vehicles in the covered road and rail loading bays. In some instances, the water sprays may also protect the structural steelwork or to enhance an escape route for Operators and Drivers.

## Ship Loading/Unloading

At terminals with river or sea-facing berths, marine loading arms are usually installed to load and unload ships. They have the advantage that high throughputs can be achieved and the disadvantages of using flexible hoses can be avoided.

Firefighting philosophy has evolved. It used to be that LNG carriers would have 'fire wires' hanging over the bow. In the event of a fire, tugs could quickly hook up to the fire wires and tow the vessel out to sea. The current philosophy is that the ship stays in the berth to provide access for land-based fire fighting teams and an escape route for the crew.

Fires on board a vessel are the responsibility of the ship's Captain. Firefighting facilities at fuel jetties are usually provided to fight fires on the shoreside, not on the tanker.

Typical systems are:

- Dry powder/twin agent to 'knock down' a small fire;
- Sprinkler systems for structural protection;
- Firewater connections to the hydrant system and foam;

In addition, there is usually a combination of manually operated and remote operated fire monitors. These may be placed so that cooling water streams may be directed onto the side of the ship, or, set to fog, to provide radiant heat protection for people and equipment on the shoreside (such as a gangway to allow the ship's crew to escape).

The only exceptions to this general rule are remote-operated fire water monitors, which are mounted at the top of the highest structures. The placement is to provide a cooling water spray onto the manifold area of the ship, regardless of the freeboard or state of the tide.

The ship's manifold is usually manned during transfers and has valves and controls to start and stop pumps. The terminal's loading arms are connected to the ship's manifold, either by a bolted flange or a quick release/no spill coupling. When a fire alarm sounds (on the ship as well as on the terminal) the ship's pumps will be shut down or the valves on the manifold will be closed to stop the transfer. Therefore, it is advisable to keep the ship's manifold area protected as long as possible.

If the flanges at the ship's manifold leak and these leaks arefeeding the fire, monitors mounted high on the jetty's structure can be used to spray foam onto the tanker's manifold area to extinguish the flames and cool the structure.

## Passive Protection

Passive protection, as a design philosophy, is usually built into the construction of the terminal. The objective is to minimise consequential damage in the event of a fire or explosion. It may be as simple as using concrete columns instead of steel as they survive longer at elevated temperatures. Or steel structures can be concrete coated (spray on concrete is often favoured). Or a steel structure can be given an intumescent coating to provide (time-limited) protection.

## Site Vehicles

Vehicles, unless specially designed or modified are likely to contain a range of potential ignition sources. Site rules should be clear about where normal road vehicles are allowed to travel.

Spark ignition vehicles are not allowed to enter Hazardous Areas. In many cases, diesel engines are allowed on site but not ATEX certified. To make them safer, diesel-powered vehicles can be fitted with spark arresters on the exhaust and over-speed protection.

·············●●●●●●●●●●●·········

# CHAPTER 5

# *Commercial Issues*

## 5.1   Business Models

### Core activities

The business of all liquid storage Terminals is to receive cargos in bulk and to break the cargo into smaller parcels to allow their distribution or onward movement.

There are three basic **business models** but a wide variety of variations:

- Terminals operating for the resale of products to smaller distributors and the downstream market. All products stored within the Terminal are owned by the Terminal Operators. Typically, this would include bitumen and petroleum products, selling to end users and petrol stations.

- Terminals leasing storage capacity to manufacturers, where specific tanks are leased for the exclusive use of a single Client. Because products cannot be comingled, the tanks, product lines, import and export facilities are dedicated to a specific Client.

- Terminals leasing storage capacity to manufacturers, where identical products are comingled within the tanks. Effectively, the Terminal operator rents space within existing tanks and does not need to worry about product contamination. Frequently used where finished products have well-established standards with little/no variation in product quality, such as gasoline and diesel.

*Note:The Terminals may provide additional services such as testing, blending, and heating which are reimbursable. Loading and unloading trucks, railway wagons, ships, or barges may involve reimbursement over and above the storage fee.*

In addition, there are niche operators, such as airport refueling companies. They are reimbursed by the Oil Companies to provide tankage, pumping and direct access to the airport refueling distribution networks. They also provide airside equipment and employ the staff who refuel the planes.

### Value Added Services

The bulk of their income is generated by the storage fees the Terminal charges for the safe storage and distribution of liquid products. However, some terminals also generate additional revenue by performing additional services for their Clients

### *Blending*

Blending is taking two or more components and mixing them so that they form a different homogenous

93

product. This is typically done when mixing various refined components to achieve a specific gasoline or diesel blend.

A few words on the historical background of gasoline blending are justified at this point.

In the days of using lead as an anti-knock additive, the octane of fuel was determined using a standard pattern single-cylinder engine (known as a 'knock engine'). Test batches of fuel were prepared in a laboratory and the test gasoline was compared to these reference fuels. But the trick of making a profit with gasoline blending is to do it so that it **meets** the specification but doesn't **exceed** it.

Anything that exceeds the legal minimum is referred to as 'giveaway'. So with constant technical evolution and under continuous pressure to make more profit, the blending operations became more sophisticated.

Originally carried out in small batches, product parcels became larger as bigger tankers became available to ship products to overseas markets. Blending became a continuous process, with 'fast loop' analyzers continuously testing to ensure the blend met the specification. Exotic hardware and software products (like the Foxboro Blendtrol) came and went.

Nowadays, dedicated gasoline blending facilities seem less popular. It is not uncommon to pump a variety of gasoline components straight into a ship's tanks, assuming the merging in the tank to be sufficient to mix the components. Part of the credit for this is due to the increasing sophistication of DCS/SCADA control systems, which will ramp up, ramp down and flush lines to very precise flow limits.

Whilst gasoline blending is mostly restricted to refineries, blending lubricating oils is more common. Because the volume of lubricating oils sold is low compared to gasoline, the blending facilities are smaller, simpler and suitable for batch operation. This suits the market, as different blends and formulations may be sold under different brand names, although all originating from the same source.

For the Client, getting the Terminal to do the blending has significant advantages:

- The cost of setting up the blending operation is no longer a Client capital expense but financed by the Terminal and repaid over the life of the contract;

- The Terminal's hardware can be used by multiple customers, bringing down the cost for individual Clients;

- A Client can serve a market without any physical presence in the country as all operations are outsourced;

- The Terminal already has the infrastructure to deal with the products, skilled staff, and a demonstrated capability to comply with any necessary legislation;

Most lubricating oils are based on a few fundamental components. The first is the base oil, which comprises 60–90% of a typical lubricating oil. There are several classes of base oil, both mineral and synthetic oils, and the product achieves its basic performance properties via the right mix.

The base oil affects the end product's volatility and stability, but also its minimum flow properties, internal friction and how well it protects against corrosion. Synthetic base oil affords broader options and enables the product to be customised in more ways.

A viscosity modifier influences the oil's thickness and viscosity even when temperatures change.

Various additives can function as detergents and anti-oxidants. As many terminals also have a laboratory on-site, they can certify product quality and provide quality assurance in-house, without the complication of external laboratories.

As far as the Terminal is concerned, the equipment necessary for lub. oil blending is straightforward and need not consist of anything more elaborate than a couple of small tanks, transfer pumps with flow metering and in-tank mixers.

## Packaging

A terminal may receive a large chemical cargo from a Client who is a supplier or manufacturer. However, the Client may wish some of that cargo to be prepared for sale to the wholesale or retail markets. Rather than investing in their own facilities, it may be more convenient to pay the terminal to perform this service on their behalf. This may require the following additional facilities:

## Bottling Line

As this isn't a core activity for the terminal, bottling lines tend to be simple with limited throughput. They are exclusively for nonfood products as the level of cleanliness for the handling of food products is seldom achievable in most Terminals.

For distribution in a retail market, the product is usually delivered in 1, 2 or 4 litre plastic 'jerry can' containers. The equipment to do this will consist of gravity filling machinery, capping machinery and labeling machinery, all connected by conveyors. Due to the low volumes, machinery can be hand feed with little in the way of automation.

Boxing and shrink wrap may also be required by the Client.

## Drums

For industrial use and wholesalers, the Client may request the liquid be supplied in drums. Drums can be between 30 and 212 litres and be made in plastic or metal, with or without a removable lid. With such large containers, the filling process is usually manual and the major problem becomes mechanical handling of heavy, but not necessarily robust, drums.

It is common for drums to be shipped on wooden pallets for easy handling by a stacker truck and shrink-wrapped.

## Warehousing

Where a terminal has blending or bottling activities, likely they have also invested in a warehouse. Blending and bottling tend to be batch operations and a consequence is that quantities of the product have to be stored until required by the end-user or wholesaler.

This building will probably have vehicle loading bays, industrial racking, pallet moving equipment, security, and maybe even fire protection facilities. Many warehousing activities are cyclic and seasonal. If space exists, it can be used by 3rd parties to store and distribute complimentary liquid products.

Many manufacturing companies concentrate on maximising production rates as a way of minimising production costs per unit. When demand for the product is low there is a need for overflow or interim storage space. The terminal can utilise its unused floor space to generate additional income without any additional capital or revenue expenditure.

## 5.2    Business Systems

I presume that in days gone by, operating a tank terminal was a relatively straightforward operation. However, as time has passed the legislative burden, plus increased Client expectations and a highly competitive market have changed the business environment.

Small terminals, frequently part of a bigger group, can avoid many of the requirements and obligations of the larger terminals. Standalone terminals don't have this luxury and are disproportionately affected by the sheer volume of paperwork and bureaucracy involved in day-to-day operations. Manpower costs are usually the biggest operating cost for terminals. As a result, how the terminal handles the essential paperwork necessary for operating can profoundly affect its commercial viability.

## A   Business Essentials

Some core activities are required regardless of the size of the terminal, its business model, or the products stored. The list below covers all topics although some (lucky) terminals may escape them.

### *Payroll*

This covers a wide range of Human Resource Department activities although purists with an HR background would argue that the payment of salaries is strictly a financial responsibility.

This will include;

- Recruitment;
- Salaries for full-time and part-time staff, employees, and contractors;
- Welfare and Medical;
- Training courses and training records;
- Pensions;
- TUPE liabilities;
- Employment statistics and records;
- HR policies and procedures;
- Compliance with Government legislation on social issues;

### *Commercial & Financial*

This will include;

- Material agreements including software licenses;
- Agreements for software support;
- Joint ventures and agreements;
- Commercial agreements with Lenders and Investors;
- Commercial agreements with Clients;
- Intercompany agreements;
- Leases and subleases;
- Property ownership;
- Audited accounts;
- Budgets and forecasts;
- Auditor's reports;

- Cash flow statements;
- Investments;
- Insurance policies;
- Records of Client billing and supporting information;
- Records of consumable and material costs;
- Records of operating expenses;
- Tax returns;

## Environmental

This may include;

- Records of waste discharge and disposal;
- Pollution control procedures and policies;
- Discharge records of GHG;
- Discharge records of VOCs;
- Wastewater treatment and discharge;
- Obnoxious smell complaints;
- Asbestos records;
- PCB records;
- Lead contamination records;
- Chlorofluorocarbon records;
- Regulatory notices;
- Past and present enforcement notices;
- Noise audits;
- Spill control plans;
- Records of spills;
- Emergency response plans;
- Environmental policies;
- Groundwater sampling and testing;

## Legal

This may include;

- Patents or copyright designs, logos, or slogans;
- Trade names;
- Government & Local Authority permits, approvals, and authorizations;

\* \* \*

## B Operating Essentials

### Operation

This may include;

- Fire detection;
- Fire alarms;
- Fire protection;
- Overfill protection (independent of TGS);
- Gas alarms;
- Emergency Shut Down (ESD);
- Tank gauging system (TGS);
- Truck and rail tanker loading/unloading management;
- SCADA, DCS, or similar
- Terminal Management System (TMS);
- Uninterrupted Power Supplies (UPS);
- Flow metering;
- Value added services;
- On-site/off-site laboratory services;

### Maintenance

- Design/construction records;
- Tank inspection policy (see 'Risk Based Inspection');
- Pipework inspection policy (see 'Risk Based Inspection');
- Maintenance planning and maintenance records system;
- Records of plant testing;
- Register of safety critical items;
- Scaffolding records and inspection policies;
- Competency register;

### Site Services

- Communications
  - ❖ Phone lines;
  - ❖ Broadband;
- Electrical
  - ❖ Connection to the national grid;
  - ❖ Backup generators;
  - ❖ Substation;
- Fuel Gas
  - ❖ Connection to the national grid;
  - ❖ Imported gas;
- Service Gas (i.e. Nitrogen)
- Heating
- Refrigeration/Cooling

- Water
  - ❖ Clean/Potable Water
  - ❖ Connection to Municipal potable water supply
  - ❖ Foul water drains
  - ❖ On-site treatment / storage
  - ❖ Connection to Municipal sewerage system
  - ❖ Contaminated (Oily) water drains
  - ❖ On-site treatment/storage
  - ❖ Storm water drains

## Security

Terminals located in ports may need perimeter security and access control systems that comply with **the International Ship & Port Security** (ISPS) requirements. This may include:

- Perimeter fencing includes anti-climb detection and protection;
- Access control for manpower and vehicles;
- Access recording for manpower and vehicle;
- CCTV;
- Escape routes and emergency procedures;

## 5.3    Accreditation

### ISO 9001 - Quality System

### Background

There are competing definitions of quality. Some people claim it has something to do with excellence. Others will define it as meeting (not exceeding) the Client's requirements.

BS 5750 (Quality Systems) started life as a procurement guide for the Ministry of Defence in the UK. In 1979 it was issued as a British Standard, and as I recall, consisted of 4 pages that nobody could understand.  Over time it was adopted and re-issued as European Standard EN29000 and as ISO 9000 in 1987.

ISO 9000 has expanded to become a family of standards on quality management and quality assurance, and many customers regard it as a prerequisite for all prospective suppliers.

Originally developed for manufacturing, its scope has now broadened to cover all types of organisations. It is accepted that the requirements of ISO 9000 will continue to evolve and expand over time. In fact, ISO 9000 has become an industry in itself. There are training companies and independent auditors plus large numbers of people employed by the Operators to handle all the additional documentation. It is a self-perpetuating cycle, with profits for everyone except the Client who ultimately pays for it.

The basic philosophy is that to provide 'a quality product or service' requires detailed procedures, and processes to ensure all issues are handled consistently and logically. In manufacturing, it is easy to see the logic. To make sure the nut and bolt sent to the Client fits, there are inspection stages throughout the manufacturing process, and the workflow through the factory is monitored and checked against established procedures. However, for a majority of companies where ISO 9000 is applied, this philosophy does **not** ensure quality, only repeatability and conformity to a standard.

To be certified as compliant to ISO 9000 an organisation's performance is monitored against its own procedures. ISO 9000 does have its detractors, as it does not apply to small organisations and, in practice, it does nothing to improve product 'quality'. However, to say so out loud is tantamount to heresy to some people, especially ISO 9000 auditors.

Originally the American market was hesitant to adopt ISO 9000 as it was seen as increasing manpower and workload with no clear financial benefit. However, over time this resistance has been worn down and now even American Defence Contractors have to be certified. There is a (justifiable) argument that what was relevant and important at the time of the Standard's conception has changed as markets and organizations have evolved, and thus it has little relevance to today's organizations.

A competing approach sometimes employed is the **Total Quality Management** (TQM) philosophy, developed in Japan during the 1950s. This relies on 13 key points and is based on people at all levels of the company being involved. By way of explanation, the points are:

- Allocate resources to provide for long-range needs rather than short-term profitability:
- Cease dependence on inspection;
- End lowest-tender contracts;
- Constantly improve the system;
- Institute training on the job;
- Institute leadership;
- Drive out fear (!);
- Break down barriers;
- Eliminate slogans and posters demanding zero defects;
- Eliminate arbitrary numerical targets;
- Permit pride of workmanship;
- Encourage education;
- Management's commitment;

## ISO 9001 in theory

Despite the disparaging remarks above regarding ISO 9000, it has become universally accepted worldwide, so anyone dealing with tank operations and terminal operations will need at least a basic understanding of how ISO 9000 works.

The ISO 9000 family contains these standards:

- ISO 9000:        Quality Management Systems - Fundamentals and Vocabulary
- ISO 9001:        Quality Management Systems - Requirements
- ISO 9004:        Quality Management - Quality of an Organization
- ISO 19011:       Guidelines for Auditing Management Systems

The 2000 revision of ISO 9001 had 5 main goals;

- Meet stakeholder needs;
- Be usable by all sizes of organizations;
- Be usable by all sectors;
- Be simple and clearly understood;
- Connect quality management system to business processes;

By the time revision 2015 had been issued, the goals had expanded. A cursory glance shows that the new ISO 9001 incorporates quite a few TQM objectives:

- Customer focus
  - ❖ Align organizational objectives with customer needs and expectations;
  - ❖ Meet customer requirements;
  - ❖ Measure customer satisfaction;
  - ❖ Manage customer relationships;
- Leadership
  - ❖ Establish a vision and direction for the organization;
  - ❖ Set challenging goals;
  - ❖ Establish trust;
- Recognize employee contributions
  - ❖ Engagement of people;
  - ❖ Make people accountable;
  - ❖ Enable participation in continual improvement;
  - ❖ Evaluate individual performance;
  - ❖ Enable learning and knowledge sharing;
- Process approach
  - ❖ Manage activities as processes;
  - ❖ Prioritize improvement opportunities;
- Improvement
  - ❖ Improve organizational performance and capabilities;
  - ❖ Empower people to make improvements;
  - ❖ Measure improvement consistently;
- Evidence-based decision making
  - ❖ Use appropriate methods to analyze data;
  - ❖ Make decisions based on analysis;
  - ❖ Balance data analysis with practical experience;
- Relationship management
  - ❖ Establish relationships considering both the short and long term;
  - ❖ Collaborate on improvement and development activities;

This is just a list of objectives but the standard doesn't define what must be done to meet any of those objectives. When trying to set up the procedures (so you can be monitored against them) a little creative thinking has to be applied. For instance, the first objective in this list is 'Customer Focus'.

To tick this box, you must set up a procedure to get feedback from the Client, and a way to ensure that feedback is assessed and ultimately reflected in the service/product. For many organisations trying to apply ISO 9001 to their business, this will not necessarily be easy. Plus, there has to be a process that says how the procedure is reviewed and who is allowed to approve it. There also have to be processes that say how current procedures are modified in the future, who is trained and how their training and performance are monitored.

Typical procedures found in Terminals and Refineries include:

- Quality Policy / Quality Manual;
- Procedure for Control of Externally Provided Products and Services;
- Record of Maintenance and Calibration;
- Staff Competence and Training Records;
- Design Records;
- Construction Records;
- Inspection Records;
- Record of Changes;
- Record of Product Characteristics;
- Evidence of Conformity;
- Record of Nonconformity;
- Monitoring Performance Information;
- Internal Audit Program and Results;
- Management Reviews;

Any institution wishing to become accredited as an ISO 9001 supplier will need to identify those activities critical to their operation. They will prepare procedures, trial them to make sure they work in practice, and audit trails constructed to justify actions and decisions.

After a period of using the new procedures, an auditor is invited in to conduct a compliance audit. If the auditors find the management's approach is acceptable, the organisation is accredited. In the following years, it will be re-assessed.

If there are any minor deficiencies, the auditors can issue a **Non-Conformance** or **Corrective Action Report** (CAR) for remedial action. In the event of a major deficiency, accreditation can be removed.

### *ISO 9001 in practice*

Despite the long pre-amble, Engineers at Terminals or Refineries can, for practical purposes, ignore ISO 9001 (unless their job is to work on its implementation). The fact that a site has already been accredited means that all the systems are in place and have been validated by an external auditor.

The procedures, method statements and processes at each site may appear to be generic or based on industry standard practice but they are specific to that site. It is the Site Operator's responsibility to train people who are new to the site, which is why many training sessions are followed by a quiz so the

Owner can prove to the auditor that the training has been carried out to an acceptable standard.

Most Owners have to employ staff specifically for the operation of ISO 9001. Systems have to be set up, audit trails created, records maintained, and staff trained (and retrained every time a procedure changes). For a standalone Terminal, this is a big and continuous investment in manpower that doesn't generate any direct economic return, but because of the market, it is an unavoidable expense.

Terminals and Refineries operating as groups can use 'group procedures' as one way of minimising manpower overheads at each site. However, for historical reasons, most Terminals, even when part of a larger group, tend to produce their own procedures and to be independently audited.

## ISO 14001 - Environmental Management

Environmental monitoring is becoming increasingly important. Not only is it the social responsibility of the Owner, but noncompliance with legislation can be financially disastrous for even the biggest institution.

Whilst ISO 9001 pretends to improve the quality of services and products (and maintains a pretense that cost benefits will follow for the adopter), ISO 14001 is more honest as it sets minimum standards for environmental monitoring and reporting. Obviously, there is a tangible cost to meeting all these standards.

At present, ISO 14001 is not generally demanded by prospective Clients and Customers, but it is inevitable that things will change, especially as the drive towards 'net zero' continues.

ISO 14001 provides a framework through which management can deliver environmental performance improvement in line with its environmental policy commitments. It specifies requirements for an organisation to proactively identify and understand the environmental aspects and impacts of its activities, products and services.

Looking through the adverts for third-party trainers and auditors for ISO 14001 is always amusing. They feel the need to claim that the standard will bring efficiency improvements and cost benefits.

Other claimed advantages include legal compliance, improved reputation, competitive advantage, and a culture of continuous improvement. In truth, cost benefits are always elusive, so any Owner seeking certification to ISO 14001 must first be clear that potential Customers will be specifying the Standard as a mandatory requirement or see indirect benefits.

From a practical perspective, the process of seeking accreditation does help focus the Owner on:

- Reduction in Pollution;
- Reduction in Waste Generation;
- Elimination of Unintended Discharges;
- Lifecycle costs;
- Energy Management;
- Utility Usage;

Like ISO 9001, reading the standard will not give any clues as to how to build systems that meet ISO 14001. The section headings in the Standard are vague and give little in the way of guidance. In some ways, this is to be expected as the Standard is written to apply to many different organisations. It also means any auditor has great flexibility when assessing the Owner's proposals.

Typical environmental procedures and documents found in Terminals include:

- Environmental Management System;
- Environmental Policy;
- Environmental Risks and Opportunities;
- Environmental Objectives;
- Operational Control Procedures;
- Procedure for Emergency Response;
- List of Interested Parties;
- List of Legal Requirements;
- Competence records;
- Evidence of Communication;
- Monitoring Performance Information;
- Compliance obligations record;
- Internal Audit Programme and Results;
- Management Reviews;
- Nonconformities and Corrective Action

The ISO 14000 family contains these standards:

- ISO 14001:          Environmental Management Systems
- ISO 14004:          Environmental Management Systems
- ISO 14005:          Environmental Management Systems

Engineers at Terminals can, for practical purposes, ignore ISO 14001 (unless their job is to work on its implementation). The fact that a site has already been accredited means that all the systems are in place and have been validated by an external auditor.

The procedures, method statements and processes at each site may appear to be generic or based on industry standard practice but they are specific to that site. It is the Site Owner's responsibility to train people who are new to the site.

Many Owners have to employ specialist environmental staff. For ISO 14001, systems have to be set up, audit trails created, records maintained, and staff trained (and retrained). For a standalone terminal this is a big and continuous investment in manpower that doesn't generate any direct economic return, but may be an unavoidable expense.

## ISO 45001 - Occupational Health

ISO 45001 sets out the minimum requirements for occupational health and safety management best practices.

As of 2018, ISO 45001 replaces the previously applicable OHSAS 18001. Companies with occupational health and safety management systems under OHSAS 18001 had three years to transition their systems to meet the requirements of the new standard.

ISO 45001 requires Owners to develop procedures for the investigation and reporting of all operational health and safety incidents and develop plans for corrective action.

Not frequently seen as a benefit to potential Customers as a dead employee at the Terminal is the Owner's responsibility, not the Customer's.

## ISO 50001 - Energy Management

Effective energy management isn't just good for business; it's also becoming a requirement. The best way to demonstrate your commitment to energy management is to be certified as being compliant with ISO 50001. The international standard outlines energy management practices that are considered to be the best.

Not frequently seen as a benefit to potential Customers as any cost savings accrue to the Owner, not the Customer.

## ISO 22000 - Food Safety Management

For those sites storing food products or components, this standard helps organizations identify and control safety hazards.

## ISO 27001 - Information Security

The standard applies to organisations that wish to assess and prevent information security risks. Not generally required by the Customer as they assume data security is absolute anyway.

## Authorised Economic Operator

AEO status is an authorisation issued by customs administrations in the European Union (EU). It certifies that the holder has met certain standards concerning compliance with customs rules and financial solvency.

For Terminals in the EU, a certain amount of international business is almost inevitable due to the proximity of borders and the globalisation of trade. For those Terminals located in the ARA (Amsterdam-Rotterdam-Antwerp) cluster, much of their international trade is carried out using river barges on the Rhine. Where terminals import products that carry duty (such as gasoline) the duty may not be payable until the product reaches the end user. Therefore, the terminals may be importing and exporting products where no duty is payable as well as fuels for domestic use where tax must be paid.

In addition, fuels due for export may be stored at the terminal long term (across accounting periods and financial years, and ownership of the product may change). As a result, tax paperwork and liabilities can be quite complex when importing and exporting these fuels. Hence some terminals in the EU need to be accredited as AEOs.

Being certified as an AEO means that customs procedures are simplified and a lot of duplicated paperwork is avoided.

**\* \* \***

## 5.4    Emissions

### Green House Gas (GHG) Emissions

At least 40 countries require facilities or companies to measure and report their emissions periodically. These reports create a data pool that is used to inform environmental policy decisions and track progress.

Since 2009, the United States has required facilities emitting at least 25,000 metric tons or more of carbon dioxide to report their greenhouse gas emissions to the Environmental Protection Agency every year.

The E.U. requires large companies to report their environmental and social impact. This is called non-financial reporting and is covered under the Non-Financial Reporting Directive, or Directive 2014/95/EU. However, currently, this only applies to larger companies.

In the United Kingdom, only incorporated companies listed on the main market of the London Stock Exchange, NYSE, or NASDAQ have to report their emissions. Those with 500 or more employees must report more information than smaller companies with fewer employees.

Since 2007, Australia has had mandatory emissions reporting under the National Greenhouse and Energy Reporting Scheme (NGERS). Only companies that meet certain emissions thresholds are required to report.

### Volatile Organic Compound (VOC) Emissions

#### *What are VOCs*

Volatile Organic Compounds VOCs) are organic compounds that vaporise (or evaporate) at normal room temperatures and pressure. Many VOCs are harmful to the environment and human health and, as such, are strictly regulated.

Common examples of VOCs that may be present in our daily lives are benzene, ethylene glycol, formaldehyde, methylene chloride, tetrachloroethylene, toluene, xylene, and butadiene.

Benzene and ethylbenzene exposure is linked with an increased risk of cancers. Toluene and xylene are non-carcinogenic, but they may produce reproductive adverse effects.

Many VOCs are of natural origin. Natural sources of VOCs include forests, termites, oceans, wetlands, tundra, and volcanoes.

#### *Sampling for VOCs*

The first step in monitoring VOCs is the collection of air samples. A Gas Chromatography is used to analyse the samples, usually at a specialist-certified laboratory.

Measurement of VOCs in ambient air is often difficult, because of the variety of VOCs of potential concern, the variety of techniques for sampling and analysis, and the lack of standardised and documented methods.

## Dealing with VOCs

Vapor Recovery Units (VRUs) are one of the most efficient ways of capturing the vapors generated from standard oil and gas production processes and generating revenue from this by-product as a result.

Vapor recovery is the process of removing harmful vapor to prevent the release of toxic pollutants into the environment. In the United States, The EPA mandates operators to carry out vapor removal in hydrocarbon storage facilities. Operators must eliminate at least 95% of vapors produced in Terminals.

VRUs come in a variety of configurations, but usually include a compressor package and a 'scrubber'. The procedure involves the following steps:

- A gas compressor sucks the hydrocarbon vapour into the scrubber;
- The scrubber removes the water, debris, and unwanted fluids;
- The recovered hydrocarbon liquid is returned to storage;

When recovery of hydrocarbon vapors is not a viable option, the vapors may be destroyed in a flare, an enclosed combustor, or a thermal oxidizer. For many applications, a simple vapor combustor (or enclosed flare) can be successfully used as a safe and economical method of controlling vapors.

## 5.5    Seveso

### Background

In 1976 there was an industrial accident in a small chemical manufacturing plant north of Milan. This resulted in a major release of dioxins, potentially affecting a large local population.

It was soon recognized that no consideration had been given to setting up any type of warning system or health-protection protocols for the local community. As a result, the local population was caught unaware when the accident happened and thus was unprepared to cope with the danger of an invisible poison.

### Consequences

Reacting to public outrage (and incorporating the lessons learned from previous incidents such as Flixborough, Bhopal, and Enshede) the EU introduced Council Directive 96/82/EC. The Directive aims to improve the safety of sites containing large quantities of dangerous substances. It was updated and revised and is now replaced by the **Seveso III** directive (2012/18/EU).

The law now covers about 12,000 industrial sites in Europe where large quantities of chemicals and petrochemical are produced or stored. There are specific requirements for Companies handling these substances above certain thresholds:

- The operator is to notify establishments where dangerous substances are present;
- The operator is to prepare a major-accident prevention policy;
- The operator of upper-tier sites is to prepare a safety report;
- The operator of upper-tier sites is to prepare an internal emergency plan;
- The authorities are to prepare an external emergency plan for upper-tier sites;

- The public is to be informed of safety measures and of the requisite behaviour in the event of an accident;

The law also:

- tightens the procedures for public consultation on projects, plans, and programmes involving plants covered by the legislation;
- ensures, through changes to land-use planning laws, that new plants are sited a safe distance away from existing ones;
- allows people to go to court if they consider they have not been properly informed or involved;
- introduces stricter inspection standards for the various installations, to ensure the safety rules are being effectively implemented;

As with all EU Regulations, the objectives of the Seveso Directive are clear, but it is less clear how the objectives are to be achieved. There are also a number of associated EU Directives that can affect how the Seveso Directive is interpreted in each member country. To add more confusion, different EU countries can decide how the Directive is incorporated within their laws. In some cases, (such as in France) the Directive has been absorbed within existing HSE legislation. In other cases, (such as the UK) new regulations were introduced and are policed by the Government's Health & Safety Executive.

In the UK these regulations are known as "Control of Major Accident Hazards" (COMAH) and numerous documents are available on the HSE website explaining how the rules are applied.

COMAH aims to prevent and mitigate the effects of major accidents involving dangerous substances which can cause serious damage/harm to people and/or the environment. COMAH treats risks to the environment as seriously as those to people.

"**Lower Tier**" sites (determined by the inventory on site) have to prepare a Major Accident Prevention Policy.

'**Top Tier**' sites have a number of additional requirements including a detailed Safety Report. Regular inspections are carried out by the Environment Agency acting jointly with the Health and Safety Executive (HSE) or the Office for Nuclear Regulation (ONR).

The implementation of COMAH/Seveso controls is an issue that is constantly under review and hotly debated. Documentation and monitoring procedures are extensive and time consuming. Any tank farm Owner who falls within the Regulations will have staff dedicated to maintaining the authorisation to operate.

## Who is Affected

In the UK, any site that stores a sufficient quantity of dangerous substances. Other European countries have variations on this theme and will have slightly different interpretations and responses, so we will use the UK COMAH regulations as a relevant example.

48 named substances determine if a site falls within the regulations. The quantity stored determines whether the site is classed as 'Upper Tier' or 'Lower Tier'.

Dangerous substances covered by the COMAH Regulations include named substances (e.g. hydrogen, ammonium nitrate, etc.), and generic categories with health, physical and environmental hazards.

All Owners must:

- notify the Environment Agency of the basic details or your operation;
- prepare a major accident prevention policy; and
- develop a safety management system;

In addition, Upper-Tier operators must:

- prepare a safety report;
- prepare and test an internal emergency plan for the site;
- supply information to the local authority for external emergency planning purposes;
- provide information to the public about the activities.

····•••••●••●••••····

# CHAPTER 6

# *Terminal Types*

## 6.1    Aviation Fuels

**Tank Design**

### *Airfields*

General aviation(GA) is defined by the International Civil Aviation Organization(ICAO) as all civil aviation aircraft operations except commercial.

At airfields that serve General Aviation and small Business jets, the turnover of fuel stocks is (comparatively) low and horizontal cylindrical tanks are frequently sufficient. These tanks tend to be mounted on concrete pedestals within a concrete bund wall. Tanks are set to fall to a 'boot' at one end so that water/contamination can be drained off. They are usually constructed from steel with a single skin. Due to the small size of the bunds, rainwater will be pumped out by the airfield's maintenance personnel and disposed of after confirmation that it is not contaminated with fuel.

In some locations where movements are few and demand for fuel is limited, tanks may have a footprint of an ISO container, with a lockable enclosure at one end where a pump and hoses are stored. This arrangement is most frequently seen at private landing strips which are not permanently manned or secured.

Airfield storage tanks are refilled by dedicated road tankers, which in turn are filled at a ground fuel depot that has access to bulk quantities of aviation fuels. As these depots are few and far between, road delivery costs can be a sizeable portion of the total fuel cost paid by plane owners.

At airfields that were originally military, there may be vertical, cylindrical ASTs. However, many of these tanks will be very old and their storage capacity will exceed current (non military) demand, meaning turnover of fuel stocks is low. Maintenance costs can be disproportionate, so we can expect these tanks to be progressively taken out of service and replaced with modern (smaller) equivalents.

### *Airports*

At large airports, such as national and international airports, tanks are typical Atmospheric Storage Tanks (ASTs).

The cleanliness of the fuel is a major concern to airline operators and passengers, so the interior of the mild steel tank is coated to prevent corrosion product contamination. The coating will extend to the roof and support steelwork. Specialist coatings resistant to aviation fuels are available such as Interline 984 by Akzo Nobel. They will typically be 2-part epoxy with a dry film thickness of 800 micron.

Before the tanks can be coated they need to be prepared by cleaning and shot blasting. As this involves erecting scaffolding inside the tank, building tanks for aviation fuels is more expensive.

Stainless steel tanks are an option, allowing the coating to be deleted, but the cost of stainless is prohibitive on very large tanks. The Defence specification 032 **"Internal Coating of Aviation Fuel Tanks"** is available to download. Although out of date it gives useful pointers.

Tank roofs are usually cone-shaped and tend to be self-supporting to simplify the internal structure (which makes applying coatings easier). Modern tanks may be fitted with geodesic dome roofs for weather protection. This type of roof can offer cost benefits but the main benefit is the reduced construction time.

Aviation fuels are lighter than water, so the tanks have 'cone down' bottoms fitted. This allows the water to be drained off the tank after it has settled, using a drain line connected to the central sump. There is usually a small window installed on the drain line close to a manually operated valve. It is obvious to the maintenance personnel whether they are draining water as it is discoloured or dirty. The liquid in the drain line turns clear when all the water and scum have been removed. The central sump also allows the removal of loose debris that has settled out at the bottom of the tank.

The drain is a small diameter line and is not used to pump the product out of the tank. The transfer pump suction is a dedicated nozzle on the shell, 600mm or so above the tank floor to avoid drawing in any contaminated fuel or loose debris at the bottom of the tank.

Two other features are popular in aviation fuel tanks.

- Frequently the Terminal Owner will have a preference for a **floating suction**. In these cases, the transfer pump suction is connected to a floating pontoon inside the tank, usually by articulated pipework to allow for liquid level changes. The logic is that only clean, uncontaminated fuel is at the very top of the tank and this is 'skimmed' off by the floating suction.

- Aviation fuel tanks also generally have an **internal floating deck**. If installed when the tank was built, it will likely be of the 'single deck' variety and constructed from steel. If a new deck is retrofitted to an existing tank, it will probably be made of aluminum. When used with a floating suction, the deck has to be of the 'non-contact' type. As with all floating decks, the purpose is to minimize the surface area of the product in contact with air in the tank, to reduce possible contamination.

Tanks should have an internal diffuser on the tank inlet to slow the velocity of incoming fuel and avoid splashing. For lightning protection, the tanks should be grounded, as should all associated pipework.

There are of course variations on these themes. There is a style of IFR that hangs from the tank roof so that the roof landing legs can be eliminated. In this way damage to the coating on the tank floor during inspection or cleaning can be avoided. Also, floating suctions can be fixed to the underside of the IFR (on a slide to allow for roof movement) eliminating the independent floats.

There does not appear to be any consensus between Owners on the optimum arrangement, and tank configurations will vary between sites, even when controlled by the same company. Sometimes the date of building the tank/tank pit will be the governing factor; sometimes it will be a change of company specification or maybe just a personal whim of the Terminal manager. Each configuration will have its advocates, who typically maintain their tank design is the optimum and every other design of tank is inferior.

Over the years I have entered quite a few aviation fuel tanks before their out-of-service inspection and was always impressed at how clean they were. In reality, this shouldn't come as a surprise as everyone involved with aviation fuel is obsessive about cleanliness and testing is both continuous and rigorous.

Cleaning an above-ground storage tank that has been used for storing AvTur usually involves only wiping the floor clean with absorbent wipes.

## Fuel Types

There are only two categories of aviation fuel, but a bewildering number of different specifications.

- **AvGas**, or aviation gasoline, is the general name for all fuels used in piston-engine aircraft. Some AvGas still contains tetra-ethyl lead additives to lubricate the engine. AvGas will normally only be found at smaller local airfields or private flying clubs with bulk storage.

  The most common AvGas is

  - ❖ 100 octane which is dyed green;
  - ❖ 100LL is the low lead version and is dyed blue;
  - ❖ 82UL (coloured purple) is a new grade but is not widely available.

- **AvTur** is the general name for all kerosene based fuel for gas turbines and turbo-jets.
  - ❖ Jet A is used in the United States ;
  - ❖ Jet A-1is used in the rest of the world;
  - ❖ JP-8 is specified and used by the U.S. military.

Military aircraft can climb faster than commercial aircraft so their fuel is blended with additives to eliminate freezing and clogging of fuel lines.

## Industry Specific Issues

### Handling

AvTur is considered to be a combustible liquid while AvGas is a volatile flammable liquid. Aviation gasoline has a flash point of approximately -46°C (-50°F), while kerosene based fuels have a minimum flash point of 38°C (100°F).

Aviation gasoline produces large volumes of vapour and is capable of forming ignitable mixtures with air even at very low temperatures. Kerosene based turbine fuel does not produce ignitable mixtures with air at normal temperatures and pressures.

Both gasoline based and kerosene based aviation fuels are classified as Class 1 products, and electrical equipment has to be selected to comply with the correct standards.

### Contamination

The cleanliness of the aviation fuel in the storage tank is critical. Nobody wants to be flying in an aircraft with contaminated fuel on board. Contamination falls into 2 categories;

## Water

The accumulation of water is almost inevitable in stored aviation fuels. Even if the fuel has low water content when delivered to the airport, there are multiple opportunities for moisture to be taken up.

Moisture can come from free water gathered in low spots in a pipeline, rainwater leaking past the seals in floating-roof tanks, and moist outside air entering the vents of tanks. Air flowing in and out of a tank when fuel is added or removed may also change the moisture content of the air in contact with fuel.

Water in aviation jet fuel can be detected in several different ways. The most common method is known as a water tablet test. This indicates whether the water in fuel is above or below 30 ppm, the maximum allowed limit.

## Particulates

Even small quantities of water in jet fuel pipelines will lead to rust developing and eventually entering the fuel. In addition, airborne particles like dust and pollen, rubber, and fabric particles can find their way into fuel. Particulates are determined by visual inspection of the fuel sample.

Kerosene based fuels are very susceptible to microbial contamination during storage in fuel tanks. Although aviation fuels are sterile when first produced, they inevitably become contaminated with microorganisms that are present in both air and water. Microorganisms found in aviation fuels include bacteria, yeasts, and fungi.

The average fuel storage tank, which contains fuel and a water bottom, provides an ideal growth environment for microorganisms. In these systems, the microorganisms thrive in the water phase of the fuel system, not in the fuel directly.

As they grow the microorganisms grow and form solid debris that effectively plugs fuel filters. Some microorganisms also produce acidic by-products that can accelerate metal corrosion.

Managing microbiological contamination is an important issue in the aviation sector. Organisations like the **International Air Transport Association** (IATA) and the **Joint Inspection Group** (JIG) provide extensive guidance on how to minimise dangerous contamination.

If aviation fuel has microbiological contamination, then there is no remedy. The fuel must be returned to a refinery for reprocessing.

## Terminal Layout

The aircraft at **General Aviation** (GA) airfields and privately owned airfields will typically be training or privately owned piston engine airplanes. A typical training plane such as a Cessna 152, has a fuel capacity of 98 litres (or 26 US gallons). A Robinson R22 helicopter, also powered by a piston engine, has a fuel capacity of 64 litres (or 16 US gallons). As fuel consumption is low, the fueling facilities will generally be small and integrated into the airfield design. There will be no aviation fuel terminal, just a tank of AvGas close to a taxiway or apron. In some cases, the fuel will be self-served from the storage tank via a meter. In other cases, there may be a refueling truck that goes to the aircraft and refuels it, operated by the airfield's full-time maintenance staff.

**Regional airports** may handle GA traffic as well as privately owned jets. Although the storage facilities may be similar to GA airfields, there will usually be two refueling trucks; one for AvTur and one for AvGas. Each truck will have dedicated hose reels, pumps, meters, and filling nozzles. To avoid putting the wrong fuel in, nozzle design and size are different for AvTur and AvGas.

FIG.611   For the private plane owner, overwing refueling may require a certain amount of agility. Best to ask if they supply ladders before turning up.

FIG.612   Sometimes the refueling truck comes to you, but it is best to phone ahead to check what hours they operate.
Photo; Losinj Airport

Rigid chassis and articulated tankers are commonplace and a capacity of 20,000 litres is typical.

As a typical private (business) jet such as a Learjet 45 has a capacity of 3,500 litres (or 930 US gallons) it is clear that fuel storage facilities will have greater capacity than GA airfields.

Most **national and international airports** will have a dedicated aviation fuel terminal on site and a hydrant refueling system.

Airports are constantly being regenerated and expanded to meet increasing passenger numbers. Historically, the fuel depot would be 'airside' but space here is especially valuable. Over the years, instead of the storage capacity at the depots expanding to cater for the increased volume of air traffic, the real estate has been taken over for use by passengers and airlines. As a consequence, the modern aviation fuel terminal at an airport is just buffering storage, with fuel being continuously supplied by pipeline from offsite terminals and refineries.

London Heathrow is typical of this situation where the consumption per day is currently (2021) 22M litres via 3 pipelines although they have only 52M litre usable storage capacity airside.

**"IATA Guidance on Airport Fuel Storage Capacity"** considers airside fuel storage capacity of international airports, complete with a worked example.

One feature common to aviation fuel terminals but seldom seen in other places is the filter units. These are typically vertically mounted pressure vessels with a baffle mid-way up the body.

Fitted to the baffle are removable filters. These filters are made from pleated paper or synthetic fibre and can be washable, disposable, or back flushable. Filters come in a variety of sizes and are designed for either solids or water contamination and in some cases both water and solids. The coalescer type of filters helps the small water droplets to combine into larger droplets so they can be stopped by the rest of the filter media.

Frequently the fuel inlet to the filter package is at the top, with the fuel outlet at the bottom. The top of the pressure vessel is bolted shut or has quick-release fittings for servicing. For filters where the flow through the filter is from the outside to the inside, staff can easily reach inside the housing and replace the old clogged filter cartridges.

These filter units are frequently installed at the end of a pipeline to ensure the fuel going into the tank is clean. However, they can just as easily be fitted to the tank export pipework.

A potential problem is static charges within the filter assembly but with correct design and operation, the risks can be mitigated.

## Distribution

A typical long-haul aircraft is the Boeing 777-300. It can carry up to 171,000 litres (or 45,000 US gallons). During a day at typical national or international airports, many aircraft of a similar size and capacity will be refueled.

Using 20,000 litre tankers to refuel such large aircraft is impractical. As a consequence, major airports have underground hydrant systems to distribute the fuel to parked aircraft.

The typical hydrant pumping system consists of a bank of pumps operating in parallel, a bank of outbound filter/separators and appropriate flow control valves. The pump and filter manifold system is provided with sampling points for fuel quality checks.

Where the hydrant transfer line enters the ground there should be an isolating valve. This should be in the form of a double block and bleed valve or double isolation valve to enable the underground lines to be tested.

The hydrant piping is sloped at a minimum of 1 in 200 to low point drains installed in the system to facilitate water removal and allow the system to be drained before maintenance. High point vents are installed for automatic venting of the system. If the hydrant is arranged as a ring main it allows for a return path to the tanks in the event fuel needs to be circulated, and also assists in the initial flushing process.

FIG.613    Commercial aircraft have under wing refueling points. All aircraft fuel tanks can be filled from a single connection. Refueling rates of up to 4,000 litres per min. are possible
Photo; Gracejet Aero

FIG.614    The Hydrant Dispenser connects to the underground hydrant using flexible hoses. After filtration and metering, the fuel goes to the aircraft's under wing connection point.
Photo; Business Insider

Where there are multiple fuel depots and hydrant systems at a single airport, the systems can be interconnected. Although more complex and costly this gives maximum operational flexibility and system redundancy. A failure of the hydrant system that stopped aircraft refueling would be extremely expensive and cause major problems for the airlines.

Control of the hydrant system is typically accomplished on a pressure and flow basis. A drop in pressure on the system will start one pump, and an increase in flow in the system will bring on additional pumps as required to meet the flow demand. The system will shut down as flow decreases and the final pump will shut down when system pressure has been satisfied. Large airport hydrant systems are frequently usually provided with computer based systems that not only control the flow of fuel but also manage the inventory.

Hydrant systems normally have cathodic protection to protect them from external corrosion. As with all underground lines, the pipe can be supplied with an external coating or wrapped during installation. Internally coated pipe may also be selected, although this does complicate fabrication and installation.

The hydrant main is routed to all the areas where aircraft can be parked and refueled. This will include the stands, apron, and parking slots adjacent to passenger terminals. At each location, a branch line comes off the top of the hydrant main and goes to a hydrant pit. The pit is there to catch spills and is covered by a waterproof load-bearing cover. Within the pit, there is the hydrant valve and an isolating maintenance valve.

When the aircraft is parked and secure, the hydrant pit cover is removed, exposing the hydrant valve. A fuel servicing vehicle (also sometimes referred to as a Hydrant Dispenser) provides the hoses to connect to the hydrant valve. The vehicle is equipped with metering, filtration, and control equipment to regulate the flow and pressure of the fuel before it enters the aircraft. A second hose connects to the aircraft's fuel inlet connections. Refueling rates of up to 4,000 litres per min. are possible.

Smaller aircraft have fueling ports on the top of the wing. However, on most commercial aircraft the fuel port is on the underside of the wing. This is known as a single-point fueling system as it can be used regardless of where the actual fuel tanks are located in the aircraft.

"IE1540" (published by the Energy Institute) gives helpful information on the design, construction, commissioning, maintenance, and testing of aviation fueling systems. In addition, the US Defense Department has a large number of standards dealing with systems and components used for aircraft refueling.

## 6.2   Ground Fuels

### Tank Design

Gasoline and Diesel are Class 1 and Class 2 products and therefore require different tank configurations and spacing to meet the relevant fire codes.

Gasoline storage tanks are typically Internal Floating Roof (IFR) tanks with a cone roof or geodesic structure. Within the tank is a full contact floating deck (usually of the 'single deck' type) with gas seals on the perimeter. The floating deck ensures the surface area of the product in contact with free air is minimal, and the seal prevents vapours from escaping into the area above the deck. The tanks are atmospheric unless there are issues specific to the location or product that requires the use of a low-pressure tank.

The style of the floor is immaterial in most cases. Tank turnover is rapid so water condensation in the tank is not a major issue and bacterial growth is not a significant corrosion problem. Contamination of the fuel is not as critical as it is with aviation fuels, so the expense of constructing 'cone down' tank bottoms is not justified. Tanks are not generally internally lined and the complication of a floating suction is not warranted.

Tank heating and insulation are not required unless the tanks are in a particularly low-temperature environment, such as the Arctic or Alaska where temperatures could be extremely low for extended periods. Even then Gasoline seldom freezes but does thicken or crystallize as some elements in the fuel begin to enter the first stage of freezing. This effect is usually delayed by the use of additives.

Diesel storage tanks are also straightforward. These will be atmospheric tanks with cone roofs. Diesel does not give off a lot of vapour so internal floating decks are not required. There is not generally any internal coating and floating suctions are not required. Again, the style of the bottom of the tank is not an issue.

Low temperatures are more of a problem with diesel fuel. The paraffin in diesel begins to stiffen at $0^{\circ}C$, leaving the fuel cloudy. The gel point (the temperature where fuel can no longer flow by gravity or be pumped) may be reached at $-18^{\circ}C$. Of course, these temperatures can be affected by chemical composition and the use of additives.

### Fuel Types

#### *Gasoline*

Americans refer to Gasoline (originally a trademark) as 'Gas'. In the UK it is called 'Petrol' as an abbreviation of Petroleum Distillate. Germans call it Benzin, the Spanish call it Gasolina, the Italians call it Benzina and the French (being French) call it 'essence'.

It is all (sort of) the same thing but it is not a single product. Refineries do not produce gasoline, they produce gasoline components that, when blended, meet the specification of gasoline in the country it is to be sold.

A simple example is 'straight cut naphtha' taken from a distillation column. At about 85 octane it isn't suitable for use in a modern car engine (although it may work in a WWII Jeep). So components produced by a Catalytic Cracker, Reformer, Alkylation or Isomerisation Unit (all of which have a standalone octane higher than naphtha) are blended with the naphtha. Then, to achieve the desired final octane, MTBE (an octane enhancer) or ethanol is added. The final product is a mix of components that, as a whole, meet the desired specification.

The part of the specification that everyone knows is 'octane' or more precisely 'octane rating number'. This is an index of a fuel's ability to resist engine knock (pre-ignition) in engines having different compression ratios. The octane rating of gasoline is not directly related to the power output of an engine. Using gasoline of higher octane than an engine is designed for cannot increase power output.

RON describes the behaviour of fuel in an engine at low temperature and speed, whilst motor octane number (MON) describes the behaviour of fuel at high temperatures and high speeds. Each country (or bloc of countries) adopts a specification for gasoline which lays down maximum and minimum figures for research octane number (RON), MON as well as flash point, sulphur content and vapour pressure.

Environmental legislation is evolving and getting stricter all the time. And it is not just the specifications that keep changing. Currently in the USA, regular gas is 87 octanes, midgrade is 89 octane and over 91 is premium. In Europe, regular is 95 while Super is usually designated 97 or 98. Whilst some countries follow US or European standards, many countries have their own specifications, sometimes specifying lower RON to minimize the production cost.

For many years fuels in the West have been unleaded. More recently there has been a move to include ethanol. Initially, the standard adopted was "E5" (5% ethanol) but now "E10" is commonplace across Europe, the US and Australia. However, some locations in the US and Europe sell "E85" gasoline/ethanol blends which containing 51% to 83% ethanol. This can only be used in specific vehicles designed for the fuel. It is not available in the UK.

However, not all the world is advancing so quickly or is so in tune with environmental demands. In parts of the old Soviet Union, low octane high sulphur fuels are still produced to keep legacy gasoline agricultural equipment working. This also provides a market for refineries that can't afford the investment necessary for modern fuels.

### Diesel

Diesel is known as DERV (Diesel Engine Road Vehicle in the UK for taxation purposes), gasoil, gasol, gaz-oil, gasolio, gasóleo, dieselolie, mazot, motorina, nafta, or just plain diesel. These fuels are used in compression ignition engines, rather than spark ignition such as used by gasoline engines. The word 'diesel' comes from the name of the inventor of the first motor vehicle using this type of engine.

Cetane (CN) is the diesel equivalent to the octane number. Fuels with lower cetane numbers have longer ignition delays, requiring more time for the fuel combustion process to be completed. Hence, higher-speed diesel engines operate more effectively with higher cetane number fuels.

Diesel fuel is more efficient than gasoline because it contains 10 percent more energy per gallon than gasoline. It's also safer than gasoline because its vapors don't explode or ignite as easily as gasoline vapors.

Most freight and delivery trucks as well as trains, buses, boats, and farm, construction, and military vehicles and light trucks have diesel engines. Diesel fuel is also used in diesel-engine generators to generate electricity, such as in remote villages. Many industrial facilities, large buildings, institutional facilities, hospitals, and electric utilities have diesel generators for backup and emergency power supply.

About 15 years ago diesel became the most popular fuel for private cars. Partly this was due to the lower taxes imposed by the Government but at the time it was being pushed as the most environmentally friendly fuel. Since then the preference of politicians has changed from achieving lower GHG (CO2) to eliminating particulate emissions. Despite the high efficiency of the diesel cycle and its potential to meet all current and proposed environmental constraints, vehicle manufacturers have been bullied into reducing the number of engines made, in anticipation of the growth of EVs.

Like gasoline, diesel is a product of atmospheric or vacuum distillation but has a lower flash point. In common with gasoline, there has been a worldwide adoption of low sulphur diesel fuels. These fuels may be marketed as "low sulphur" or "city" (they are the same fuel). In most of Europe, this is the only grade of diesel available at filling stations for cars and lorries.

In the UK road vehicles use (EN590) low sulphur content diesel fuel. This specification is used all over Europe. BS 2869 is a specification for industrial uses of Gas Oil-an alternative name for diesel. Grade A2 refers to diesel for off-road engines and machinery. Grade D is the diesel fuel for industrial heating and power generation. Gas Oil for home heating is more properly known as Grade C2 Kerosene.

Also in the UK, 'Red' diesel is available for specific users. The fuel used by some industries is taxed at a lower rate than the standard 'white' DERV used by cars and lorries. They include:

- Commercial vessels such as fishing and ferries;
- Agriculture & forestry;
- Rail transport;
- Heating and power generation;

Now entering the market in some countries are biodiesel fuels. '1st generation' biodiesel fuels were produced from oil crops such as rape, palm or soy, but concerns exist over their sustainability.

Looking into the future, FAME (Fatty Acid Methyl Ester) diesel may become more popular. FAME has physical properties similar to those of conventional diesel. It is also non-toxic and biodegradable. The most common grade of biodiesel in the US is "B5", a blend of 5% plant-based biodiesel and regular diesel.

HVO stands for hydro-treated vegetable oil and is a paraffinic diesel fuel. It is made from certified 100% waste vegetable oils and animal fats and offers up to a 90% reduction in net CO2 emissions. Because it meets EN 15940 standards and Fuel Quality Directive 2009/30/EC Annex II, it can be used as a direct drop-in alternative for diesel with little or no modification needed to the engines. At present (2021) the cost of HVO is 10-15% higher than diesel. User acceptance is low in many European countries and it is not just the price premium that is restricting uptake.

In California, the use of a specific grade of diesel became mandatory to reduce vehicle pollution. This is known as CARB (California Air Resources Board) diesel. Lots of things introduced in California have a habit of spreading into adjoining states but their current use appears to be limited to the west coast.

## Industry Specific Issues

The 'big' issue for ground fuels is their threatened extinction because of their impact on the environment and health.

Governments around the world have expressed their intention to move rapidly to electric-powered vehicles (EVs) and eliminate all hydrocarbon fuels.

As usual with politicians, the message doesn't bear detailed examination. A frequently quoted mantra is **"Zero Carbon by 2050"**. In practice, they don't mean carbon, they are confusing it with $CO^2$. Even this is wrong as they should be referring to GHG (greenhouse gases) not carbon. 'Zero' is always going to be achievable only with extreme costs and the year 2050 is a political target, not one suggested by the science.

Most Governments don't seem to have a detailed plan to achieve their objective. Instead, they frequently chose the route of banning something and leaving others (frequently Engineers) to solve the practical problems they are creating. Moving away from coal, gas, and nuclear to solar and wind will leave most countries short of electrical power. Small EVs are (apparently) unachievable and the capacity to mine the quantity of ingredients to make their batteries does not currently exist. The practicality and cost of blending hydrogen with natural gas for home heating and industrial processes is a problem that doesn't currently have a solution.

One of the effects of the current political environment is the impact on investment. Many corporate investors are selecting investments based on ESG criteria (environment, social, and governance). As a consequence, it has become almost impossible to find investors for any projects involving hydrocarbons. For the developed nations, this will cause enough problems but in many parts of the world, this is a pending disaster. Underdeveloped countries were relying on oil and gas to lift them out of poverty, and solve the problems of starvation, clean drinking water, and disease.

So the hydrocarbon industry finds itself caught in the middle of a financial pincer attack. On the one hand, they are investing their money to make, transport, and store the 'environmentally friendly' fuels now being demanded. At the same time, Investors have decided there is no future in oil & gas and would prefer to put their money into 'sustainable' and 'green' projects with a more secure return on the capital employed.

Whilst demand for gasoline has stabilized and in some cases has dropped, this is because of the increased efficiency of modern engines and car sales topping out. An irony is that biofuels give lower mileage than conventional fuels-this is an increased cost to consumers that appears to have been overlooked. At the same time, diesel engines, which give better efficiency than gasoline-powered engines and lower GHG emissions, are being killed off as a precursor to killing off the gasoline engine.

Looking at the current situation dispassionately, it is clear the long-term future of hydrocarbon ground fuels is not rosy, although some sustainable fuels may be developed in the future. If sales of ICE cars are banned by 2030, the majority of cars on the road will still be fueled by gasoline. Presumably, these vehicles will still be on the roads (and requiring fuel) until 2050 and beyond. At the moment there is no real alternative to diesel fuels for heavy vehicles and trucks so presumably, these will still be on sale until 2050. There is no alternative to diesel (Gas Oil) for home heating. There is no alternative to Grade D Gas Oil for power generation and industrial heating.

So presumably, the future of bulk storage facilities for gasoline and diesel, and their sustainable successor products, looks reasonably assured for the immediate future.

## Terminal Layout

When a site has both fuels, it is traditional for the Class 1 tanks to be in a dedicated tank pit, and Class 2 tanks to be in a separate tank pit. Sometimes this will be because the tank pits were constructed (or rebuilt) at different times to different standards. Or maybe the Terminal Operator built facilities for a different customer. At many sites, history has been the driving factor for the layout, not current technical or legislative requirements.

For sites storing multiple grades of gasoline and diesel, it is not unusual to have all tanks built to a common standard. The idea is that as consumer demand changes over time, some of the tanks will end up being used to store products they were not originally intended for.

Some refineries produce ground fuels and act as a wholesaler to the filling stations. In these cases, the tanks and tanker loading bays may be part of the refinery 'offsites' area. Alternatively, the ground fuel terminal may be remote from the refinery but supplied by pipeline or railway trucks. Regardless to whether the site has storage tanks and how stocks are replenished; the common feature is they will all have a truck loading facility, where the individual road or rail tanks are filled before being moved to the retail sites.

Modern ground fuel terminals will also have vapour recovery systems or vapor incinerators to limit their VOC discharges to within permitted limits.

## Distribution

The supplies reaching the Filling Stations will always get there by road tanker. For clarity:

- Refineries are the **manufacturers** of the fuel. Fuel can easily be produced to meet national specifications so consignments of fuel are shipped internationally as demand and price dictate. The international market is driven by seasonal demands (e.g. as demand for road diesel declines at the end of summer, demand for heating oil increases as autumn approaches). Most refineries can supply the local retail market direct. In some cases, the road tankers are filled from the Tank Farm in the refinery. In some cases, there will be a separate terminal outside the refinery where the road tankers are filled via a pipeline from the refinery;

- We can regard ground fuel terminals as **wholesalers**. They are frequently operated by the same oil company that supplies them with the fuel. It is not uncommon for the terminal to supply any independently owned forecourt operator as well as IOC marketing groups. The petrol you buy from a Shell Filling Station may have been produced by the local Esso refinery;

- Filling stations (where the end customer fills up their vehicle) are the **retailers**. Now frequently paired with a standalone convenience store for increased profit;

Ground fuel terminals come in many shapes and sizes. It is perhaps useful to look at the simplest operations and then see what can be improved and why.

A typical terminal in rural Africa will consist of a couple of small tanks, usually horizontal, to store the products. There will be a small transfer pump, with a spill-back loop to avoid damage to the pump. The pump sends the product to the road tanker parking area, with a simple flow meter in the line for inventory control. A rubberized fabric hose is used to direct the product into the top hatch of the road tanker. When filled, the road tanker travels to the filling stations on its delivery route.

At each filling station or forecourt, it will use a rubber hose to discharge to the tanks by gravity.

As crude, as it sounds, this type of loading operation is daily practice across hundreds (if not thousands) of small ground fuel depots. Even to the casual observer improvements will be obvious.

FIG.621   The configuration is fairly basic but in the hands of skilled Operators it works well-just not as quickly or as safely as a computer-controlled system.

FIG.622   A basic loading arrangement with pump, strainer, EOV and flow meter discharging to the top loading arm.

A typical road tanker in Africa will be a rigid chassis to survive the poor roads and the need to go off-road occasionally. Tanks are typically between 5,000 and 15,000 litres and either a single tank or several compartments. Having several compartments means that different products can be carried. But top loading (i.e. loading through the top hatch) cannot be carried out quickly as static electricity can build up. The whole filling process is timeconsuming and labour intensive (although in rural Africa this wouldn't normally be regarded as a problem).

By comparison, a typical (big) road tanker in Europe will have 5 compartments and a total capacity of up to 44,000 litres. Each compartment can be filled independently from hoses connected to the bottom of the compartment. Filling from the bottom can be done quicker (up to 2,000 litres/min.) and all compartments can be filled simultaneously. The hatches on the top of the tanks allow extraction hoods to be fitted. As the tanks are filled, the vapour in the tank is displaced and can be extracted and either recovered or disposed of.

As European regulations now include vapour recovery, the new road tankers have a hose connection to extract the vapour from the Filling Station's underground tanks. The tractor unit is articulated and detachable, increasing flexibility and maneuverability.

The changes to the road tankers improve operator safety and efficiency. The driver no longer has to climb onto the tank to open the hatches or handle heavy hoses while on top of the tank. As the tanks can be filled simultaneously the vehicle filling time can be reduced to 10-15 minutes and a range of products can be transported.

These upgraded fuel tankers could be employed in Africa but it would require a massive investment in new equipment and operator training. It is also questionable whether they would (on their own) be of any great benefit. Small road tankers are more practical given the social conditions and the roads.

The other big improvement with the distribution of ground fuels is the loading equipment at the Terminal. The Terminals have invested heavily to reduce the vehicle turnaround time and also to make the whole process safer for the operators and more environmentally friendly. This has been driven by the economics of operating these vehicles as well as more onerous legislation.

To reduce the turnaround time, each loading bay can have 5 or more bottom loading arms. These loading arms are rigid pipes with articulated joints. They are either counterbalanced or have balance springs to make them effortless to connect to the tanker using a quick-release coupling. The driver who loads the tanker (no external assistance is required) has only to connect the loading arms to the road tanker. The operation of the pumps is all automatic and controlled by a local computer.

As many of the loading bays at older Terminals were converted from top loading to bottom loading, the old top loading hoses were retained, although they are seldom used.

The driver seldom needs to work on top of the tanker. The loading bay will usually have a free-standing steel gantry with access platforms and handrails. The driver only has to walk up the staircase, open the hatches and move a counterbalanced extraction hood (to collect displaced vapour from the tank). In most cases, the driver will wear a fall prevention harness as well.

In case of spills, the road tanker is parked on a concrete apron, with a closed drain system with sufficient capacity to hold even major spills. To prevent sparks caused by static electricity, the tractor unit and tanker are earthed, and the earth line is connected to the computer control system. If the earth is removed the transfer pumps will automatically shut down.

As static electricity can be a primary cause of sparks, it is worth considering the three clear potential sources. If 'splash loading' (i.e. top loading with a hand hose), the violent movement of liquid within the tank can build up a static charge. Any contact with an unearthed metal fitting, such as a metal nozzle touching the hatch cover, can generate a spark large enough to ignite vapours. In the same situation an unearthed hose can create a static charge due to the velocity of liquid through it. Again, any contact between a metal nozzle with an unearthed tanker can generate a spark suffient to cause an explosion and fire. This is why hoses for handling powders and flammable liquids with low conductivity have an earth cable built in. The last (and least likely cause) is if the Operator builds up a static charge on his clothing.

To avoid these potentially fatal misshapes, splash loading should be avoided if possible, but if carried out, earthed hoses should be used connected to a secure and tested earth point (usually a permanent earthing rod in a concrete pit) and the Operator should wear approve PPE. The road tankers should be electrically 'bonded'to the chassis, the cab and engine and then connected to the earth at the loading bay.

Three standards, in particular, provide clear guidance on what precautions should be taken. NFPA 77, API RP 2003 and CLCTR: 50404 state that grounding (earthing) of the road tanker should be the first procedure carried out in the transfer process. There are numerous simple and cheap devices that continuously monitor and verify the grounding point. The 'Earth-Rite' RTR (Road Tester Recognition) device is typical and can verify the integrity of the earthing system and the vehicle earthing before loading commences.

Frequently the loading point gantry will provide the secure earth point, but earthing cables and clamps should be designed to connect onto earthing points on the vehicles that may have paint or corrosion. A point seldom considered is that low temperatures can increase ground resistance at the earthing rod, so the integrity of the earth should be tested regularly.

The flow control computers are central to the optimization of the loading process. As each driver parks at their loading bay, they identify themselves and the vehicle to the computer. The computer has been pre-programmed with the details of the tanker's next load and will have lined up the valves to ensure the correct products are available at the loading arm.

When the loading operation is initiated, discharge rates are chosen to ensure each compartment is filled in the optimum time. At the end of the loading process, delivery notes and waybills, even routing instructions are printed out for the driver, who can then disconnect the hoses from the tanker's quick release, no spill couplings. The driver is now clear to deliver the consignments to the Filling Stations or Forecourts.

Biofuels can deteriorate when stored for any length of time and are seldom stored in bulk. Instead, the ethanol can be blended with the regular gasoline as it is pumped into the road tanker. The correct ratios of the product are all controlled by the flow meters and the flow control computer.

## Blending

For simplicity, the following description of the blending process at the loading rack specifically refers to ethanol and Gasoline, but the process of blending biodiesel is the same.

The objective in ethanol blending is to mix precisely measured quantities of ethanol and gasoline together. There are currently two main ways to blend products during the loading process at the loading rack.

**Sequential blending**-gasoline and ethanol are individually measured through the same single meter. To accomplish this, each component of the blend is loaded sequentially, one at a time, until all blend components are dispensed. This method requires a single meter and control valve, and relies on subsequent mixing (for example in a tanker compartment) to effect the proper blend. Only recommended for older installations, simple blends or where there are fewer vehicle movements.

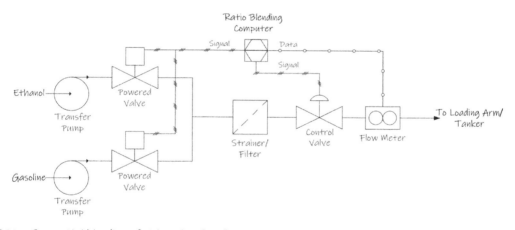

DIAG.621    Sequential blending of Ethanol & Gasoline

**Ratio-blending**-each component of the blend flows through its own individual meter and is controlled by a dedicated flow control valve. This method requires one meter and one control valve per blend component. Ratio blending is the recommended option for ethanol and biodiesel blending but is more expensive to implement and more complex to operate. However, it is quicker and usually pays for itself by reducing tanker turnaround times.

There are two alternative methods of using the same basic equipment configuration.

- **Proportional blending** is accomplished by controlling the ratio between the gasoline and ethanol at all times so the ratio of the blend is correct at all times. Therefore, the delivery can be stopped at anytime and the delivered blend will be within tolerance.

125

- **Non-proportional blending** is accomplished by delivering the ethanol at a high flow rate, so that the flow meter is operating at maximum measurement accuracy. After the ethanol has been delivered the gasoline will continue to flow until the correct final blend ratio is attained after subsequent mixing in the tanker.

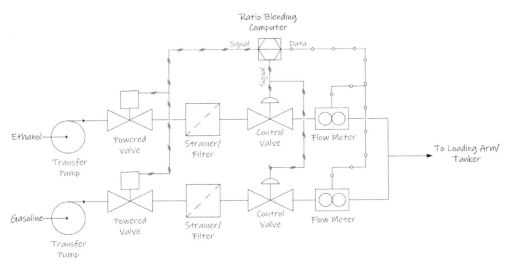

DIAG.622   Ratio blending of Ethanol & Gasoline

There are several variations and hybrids, such as Side Stream and Wild Stream which have their own advantages and disadvantages.

In addition to blending ethanol with gasoline, there may also be a need to blend different grades of gasoline to produce an intermediate grade. Some oil companies also use proprietary additives for their premium grades for marketing purposes.

Flow rates can be ramped up at the beginning of the blending operation and ramped down as the batch quantity is reached. It is also standard practice to flush the ethanol lines with gasoline after blending at some locations (ethanol is basically a solvent and has an affinity with water). To make calculations even more confusing, mixing gasoline and ethanol together causes a chemical reaction that can cause a temperature drop and the density of the blend to change!

The blending operations can be quite complex so many Terminals use dedicated ratio blending computers (such as the **FMC Accuload**). Blending computers come in a variety of configurations. Some basic models are designed for simple sequential blending, whilst others are more complex and part of a modular loading management system.

Retrofitting old Terminals with new equipment is not necessarily a problem as many manufacturers produce 'blending skids', These are complete skid-mounted packages including, if required, tanker loading arms.

All that is required at site is a straightforward hook-up to existing pipe work and power.

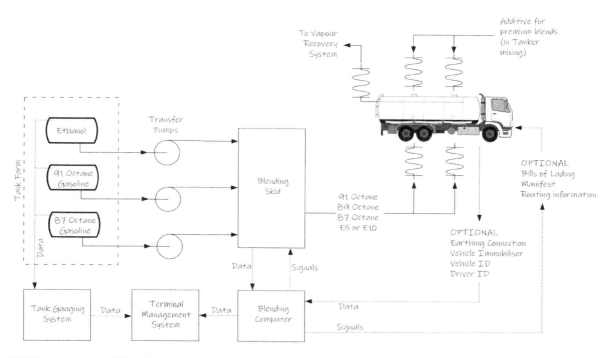

DIAG.623   Typical blending arrangement for Gasoline at a Terminal. Biodiesel blending is similar.

## Metering

Traditionally, all metering during the loading process (and unloading process) is carried out using positive displacement flow meters. PD meters measure the volumetric flow rate, but with temperature compensation, mass flow rate can be accurately determined.

There are a variety of designs, but the difference is immaterial unless you are an Instrument Engineer. Their advantage is that they are simple, robust, have a low pressure drop and the accuracy is not affected by a low flow rate. As they are powered by the flow of liquid going through the meter they do not need any power supply. Although they will tolerate medium and high viscosity, they are sensitive to impurities, but this constraint is not relevant for the blending of ground fuels.

When used as part of a blending package, the flow meter will have some form of transmitter to send signals to the blending computer, but on older installations it is common to see a PD meter as a standalone meter with mechanical totalized readout (see Fig 622). For loading road tankers using these 'basic' facilities, the Operator/Driver will usually have a secondary counter they can reset to zero to measure how much product is going into each compartment. When the permitted quantity has been delivered the Operator/Driver shuts off the flow with a manual valve or EOV.

## Tanker Equipment

There is no such thing as a 'standard' road tanker. In many countries, tankers come in a variety of shapes and sizes, most of them 'top' loading. Whilst the tank is not leaking and the tanker's engine works, there is little incentive for Owners to upgrade to more modern equipment. As a result, the makeup of the 'legacy' fleet in places like parts of sub Saharan Africa will remain unchanged for the foreseeable future. For the Operators in these Terminals, bottom loading hoses, articulated and counterbalanced arms and blending computers are not something they will ever have to worry about.

Regardless of what equipment is used to load them, tankers still have to deliver to the customers. Simple (old) road tankers have valves on the bottom of the tanks (often with buckets hanging under them to catch the drips from leaky valves). When the tanker gets to the Customer the cargo is usually pumped to the Customer's tanks. This is usually done with a power take-off from the road tanker engine to a small onboard pump, which is why the tanker's motors are kept running while offloading.

In countries with newer delivery tankers, capacities of up to 44,000 liters are the norm, and the equipment installed on them can be quite sophisticated.

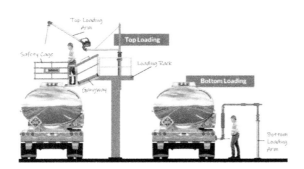

FIG.623 The difference between top and bottom loading. Bottom loading is quicker and safer, but you still need a safety cage to get access to the top hatches for vapour recovery.

FIG.624   A modern tanker with 5 compartments, bottom loading and flow computer to monitor deliveries. Tankers carry their own lightweight hoses for vapour recovery.
Photo; Alpeco

Modern practice is that the road tanker is equipped with a meter to measure the product delivered to the customers. For those vehicles that use electronic meters, flow computers can be incorporated. Frequently built into ATEX certified enclosures, these computers keep track of volumes and grades delivered and safety related information. This data is downloaded to the Terminal on their next visit. The most modern devices can control pumping operations, handle product returns and ticket printing. With this modern equipment, communications at the Terminal can be handled by Wi-Fi, Bluetooth or USB. A typical modern road tanker valve manifold and flow meter is shown in Fig. 624.

Most tankers carry hoses to connect the tanker to the delivery points at the filling stations. Typically, for bottom loading and vapour recovery, these hoses have a cam lock style connector to a 4" API valve (API RP 1004) that incorporates a sight glass to check for retained product. The design of the valve ensures quick disconnect and no (minimal) product spill.

With some products, the tanker can be unloaded by applying air or nitrogen pressure above the liquid or powders. However, this is not a common practice and pumps are frequently used. Positive displacement pumps are used for viscous products but most ground fuel tankers use a self priming centrifugal pump, usually with a magnet drive to minimize leaks.

* * *

## 6.3    Marine Fuels

### Tank Design

The SOLAS recommendation is for all commercial marine fuels to have a flashpoint above 60°C. As a result, all tanks at marine fuel terminals are built to hold Class II or Class III products.

The driving factor for tank capacity is not the volume of product sold, it is dependent on the parcel size of deliveries. As a consequence, tanks with a volume of about 10,000m or smaller are typical. Due to the tank locations, smaller tanks also give more flexibility in terminal layout. This is especially important as quayside locations are usually congested and space is at a premium.

Tanks would typically be conventional ASTs, fabricated from carbon steel. A flat roof would suffice on smaller tanks, with a cone roof on larger tanks. Tank bottoms are usually flat.

HFO requires heating so tanks to store this may have steam coils or an external heater and some method of mixing the contents. Insulation to the tanks reduces the operational cost and also inhibits corrosion of the steel tank in an aggressive marine environment.

Tanks containing distillate fuels (or blends of distillate and residual fuels) do not require heating, mixing, or insulation. The products are classified as combustible, rather than flammable, and only limited firefighting facilities are required. Electrical equipment isn't required to cope with high vapour environments so blast-proof equipment is not mandatory. However, as the tanks are usually close to the edge of the quay, secondary containment bunding capacity and integrity is essential if spills into the harbour are to be avoided.

### Fuel Types

#### *Recreational Vessels*

Most recently built recreational vessels use diesel or gasoline as engine fuel. Diesel is considered safer than gasoline for inboard engines because of its comparatively low volatility.

Many marinas and boatyards will supply fuel for vessels as this is a valuable income stream as well as a service to their customers. The refueling is generally carried out from a moored pontoon with storage tanks built into the hull, with hoses and meters on the deck.

In the UK, diesel fuel is also called Gas Oil when used for industrial purposes. One confusing issue until recently has been the status and use of 'red' diesel. Red diesel is the same as regular diesel, but sold specifically for use with machinery and engines where Fuel Duty is lower (VAT may also be lower). Red diesel has a red marker (dye) added to it to prevent illegal use on road-going vehicles.

To push the industry towards alternative low-emission fuels, from April 2022 Red Diesel will only be sold to the commercial boating industry, including fishing, inland water freight and passenger ferries. As a consequence of the rule change, there will be no sales of Red diesel for recreational use.

Depending on the local regulations, the octane rating of the Gasoline may vary (fuels are invariably sourced from Ground Fuel terminals). As the move to biofuel continues, more marinas will stock Gasoline with ethanol. E5 may be the most common but in some markets E10, E15 and E85 may be available. Boat owners have to be cautious about using fuel with high levels of ethanol-many boat engines and fuel systems were not designed for it and it tends to absorb moisture, leading to accelerated corrosion.

Fuels for outboard engines are available at marinas. They tend to be pre-blended 25:1 or 50:1 fuel/oil for 2 strokes and sold in retail packaging.

FIG.631 Refuelling pontoon in a small marina. As most recreational boats use either petrol or gasoline, there is only need for 2 pumps

FIG.632　If you own a big enough boat, they will come to you!
Photo; Maritime Bunkering Ltd

### Commercial Vessels

To push industry towards alternative low emission fuels and to maintain compliance with EU Directives, from April 2022 Red Diesel will only be sold to the commercial boating industry, including fishing, inland water freight and passenger ferries. Previously it had been allowed to use Red Diesel for propulsion of private yachts and motor vessels.

Red diesel complies with BS 2869 Grade A2 specification or EN590:2009. Smaller commercial vessels in Europe, such as trawlers and ferries, tend to use either (red) diesel or Marine Gas Oil (MGO). MGO is similar to diesel fuel but has a higher density.

IMO 2020 has transformed marine fuel from essentially a waste product sold at a discount into one of the industry's most valuable assets. They have introduced a 'sulphur' cap and ships should only use marine fuel <0.5% sulphur content.

Areas off the coast of the USA, the North Sea, and Baltic Sea have been designated as **Emission Control Areas**. Ships operating in ECAs must use fuel with a sulphur content of <0.1% and HFO is now banned in vessels operating in the Arctic.

Marine fuels are either residual products, distillate products or blends of both. There are 4 main headings, with sub-groupings defined by their sulphur content. There was an attempt by ISO to standardize product categories (ISO 8216 & 8217) but the designations used are even more confusing.

Terms such as 'low sulphur' (LS) and 'ultra low' (ULS) are used to describe generic products that meet regulatory requirements. 'LS' tends to mean products that are above 0.10% but meet a 0.50% sulphur limit. 'ULS' tends to mean a product with a maximum of 0.10% sulphur content

The descriptions below show US categories, but UK and European equivalents also exist.

- **HFO (Heavy Fuel Oil)**     Also known as *'bunker fuel'* or *'residual fuel oil'*. Until recently, most large commercial vessels used this due to its low cost. Tar-like consistency requires heating during storage. Sales of product will reduce significantly due to the IMO rules but continued use is still permissible in vessels with engine exhaust scrubbers.

- **MGO (Marine Gas Oil)**     Made from distillate only. Similar to diesel but has a higher density. The most preferred 'clean' fuel for large vessels as it has a sulphur content between 0.1% and 1.5%. Does not require heating during storage. Roughly equivalent to #2 Fuel Oil in the US.

- **MDO (Marine Diesel Oil)**   A blend of distillates and heavy fuel oil, generally with a very low HFO content. MDO is roughly equivalent to #3 Fuel Oil. Intermediate Fuel Oil (IFO) is a marine diesel with higher proportions of HFO. IFO is roughly equivalent to #4 Fuel Oil in the US.

- **MFO (Marine Fuel Oil)**     A blend of distillates and heavy fuel oil. An alternative name for #6 HFO

For legal purposes in Europe:

- "marine diesel oil" means any marine fuel which has a viscosity or density falling within the ranges of viscosity or density defined for DMB or DMC grades in Table I of ISO 8217 (2005);

- "marine gas oil" means any marine fuel which has a viscosity or density falling within the ranges of viscosity or density defined for DMX or DMA grades in Table I of ISO 8217 (2005);

There are two ways to make low sulphur marine fuels. The first is to distill the crude oil and then remove the sulphur. The second is to distill only low sulphur crude oil, in which case there is little or no sulphur to remove. However, sources of low sulphur crude are limited and we can expect prices to increase as demand for low sulphur marine fuels increases.

It is important to note that whilst all products can be used to propel the vessel, there may be other equipment on board that needs a different fuel. Ships have generators that may require diesel or LNG to operate efficiently. Where offshore drilling rigs or production platforms are located, rig support vessels will be expected to supply them with a range of marine fuels and lubricants in bulk in addition to standard fuel blends.

A more recent development is the rise in the use of LNG as a marine fuel. However, before it can become widespread more vessels have to be built with engines suitable for LNG, and more refueling terminals have to be built worldwide.

## Industry Specific Issues

### Pollution

A cursory search of the internet will reveal that HFO fuel has been widely condemned due to its adverse effect on human health. It is a combination of pollution and the sheer quantity of fuel burnt worldwide. This has prompted the move towards low sulphur versions of the distillate fuels being adopted.

All marine fuels are, by nature, high flash and low vapour products. However, due to the large volumes of product being moved from the tank to the ship (and vice versa), larger terminals will have a vapour recovery system or incinerator.

## Contamination

All gas oil or Diesel fuel in storage tanks contains or gains a degree of water content, either as free water or as absorbed water. Usually, such absorbed water content is a very small proportion.

Long-term storage may cause the fuel's properties to deteriorate. FAME type bio-diesel has the scope for accumulating a higher water content than standard diesel fuel. Significant water content may collect at the bottom of a storage tank and should be drained off at regular intervals.

However, lower levels of water content may also remain in suspension in the fuel and is worse for bio-diesel mix fuels. Micro-bacterial contamination tends to be most prevalent at the fuel/water interface.

In most cases the turnover of fuel is enough to stop this from becoming a problem, but in small marinas and boatyards, they may not sell adequate quantities of diesel, especially over the winter.

## Safety

Working in a marine environment always brings increased risk for Terminal operations and maintenance personnel. However, safety systems are well established to mitigate any additional risks to the people working on the Terminal or on the lighters.

## Spills

As the fuel is stored near the water's edge, spills will always be a concern for Terminal Operators. Oil spills can be very harmful to marine birds and mammals, and also can harm fish and shellfish. Oil destroys the insulating ability of fur-bearing mammals, such as sea otters, and the water-repelling abilities of a bird's feathers, exposing them to the elements.

Terminal operators make provisions to deal with onshore spills and cleanup. These plans will include;

- Shoreline flushing/washing;
- Vacuum trucks;
- Sorbents;
- Manual removal;
- Mechanical removal;

However, when the spill is into the water, cleanups become a lot more expensive and damaging. Especially to the Terminal Owner's reputation. When this happens, the Operator's response is twofold: containment and cleanup.

Booms are floating physical barriers, made of plastic, metal or other materials, which slow the spread of oil and keep it contained. A boom may be placed around a tanker that is leaking oil, to collect the oil, or along a sensitive coastal area to prevent oil from reaching it. In some cases, the oil can be removed using skimmers from the surface of the water so it can be collected for disposal.

More controversial methods may be used after consultation with local authorities and spill response experts:

- In situ burning, usually while it's floating on the water;
- Dispersants to minimize the oil reaching coastal wetlands, beaches, tidal flats, etc.;
- Biological agents to help break down oil into its chemical constituents;

Some spills evaporate rapidly off the water surface without any active clean-up. However, residual fuels with a high density tend to 'blob' and wash up on the local shore and coastline, requiring significant quantities of labour to manually clean the beaches and rocks.

## Terminal Layout &Distribution

There is no such thing as a standard marine fuel terminal as all are different to suit the location, throughput, and customer base. In some cases, there may not even be a marine fuel terminal as such, with marine bunkering firms buying fuel directly from the refineries and major stockists.

Small harbours have limited berthing and tend to support local fishing fleets. Tankage may consist of a few small tanks on the quayside, supplied either by rail or road tanker.

If large enough, they may import products by barge or tanker. Transfers are usually by hose from the barge to the tanks on shore as the quantities are small and there is insufficient space on the quay for a dedicated berth for tankers and permanent unloading equipment.

The demand for fuel is so limited in some harbours that road delivery tankers will drive along the quay and pump straight into the ship's tanks.

Probably the best way to explain the range of options is to look at the port of Rotterdam. Rotterdam, together with Antwerp and Amsterdam is collectively known as the ARA cluster. These three ports are one of the largest concentrations of heavy commercial shipping in the world.

When flying over Rotterdam the initial impression is the whole city consists of storage tanks and greenhouses. Rotterdam has 5 oil refineries dotted around the port area. As a consequence, bunkering companies frequently don't need their own storage tanks as the product can be sourced directly from local refineries.

Where bunkering companies have their own tanks, it can be because they buy a variety of marine fuel components and blend these to produce marine fuels to meet international standards and their Client's needs.

I know of one large 'general' terminal that has a couple of distillation columns that are used to produce low sulphur marine fuels from the low sulphur crude they import, undercutting the oil majors. This company has a long-term supply contract with a large shipping chain so all parties benefit from a secure supply chain and demand that does not fluctuate.

The price of marine fuels in Rotterdam is generally considered to undercut most other locations. Many ships are routed specifically to take advantage of the low prices here. Due to the significant demand, and active promotion by the Port Authorities, there are a large number of suppliers who provide fuels to ships as well as brokerage and fuel trading services.

Ships have many ways of being refueled. The port of Rotterdam is constantly expanding west into the North Sea, and the new quays contain fuel bunkering hydrant lines. As a result, ships can be refueled by hose from the quayside, whilst unloading their cargo. Alternatively, double-hulled barges and refueling ships, carrying product straight from the producers and refineries, can tie up alongside the ship and pump Ship to ship using large flexible hoses. Refueling whilst unloading allows the ships to offload their cargo within 24 hours and eliminates demurrage costs.

Ships moored awaiting a berth in the port can also be refueled. They tie up to buoys or dolphins in the river, the refueling ship comes alongside and a flexible hose is winched aboard (see Fig. 634).

It is not only large cargo ships that benefit from the infrastructure in Holland. Barges of all sizes transship cargoes in Rotterdam to export to the European hinterland via the Rhine. Most terminals with a river frontage in Rotterdam will have berthing for river barges and coasters, all of which require fuel.

Admittedly, having so many refineries in a small area does give Rotterdam a significant commercial advantage, but the operation in Rotterdam demonstrates how a major shipping hub can be developed when there is a desire to succeed and has Government backing.

There has recently been a dramatic scaling up of ship-to-ship LNG bunkering but worldwide the number of LNG bunkering facilities is still extremely small.

In early 2019 there were just six LNG bunkering vessels around the world. As of January 2020, there are 12 in operation with a further 27 on order and/or undergoing commissioning, the majority due to come into service within the next few years. Rotterdam already has a dedicated refueling berth where ships using LNG for propulsion get priority.

## 6.4 Asphalt & Bitumen

### Tank Design

The bitumen storage business is difficult to categorise as it covers a broad spectrum of users and businesses. Whilst the majority of bitumen is used as a binder for asphalt for roads, it is also an important ingredient in many other manufacturing industries. As a result, volumes stored on site can vary widely.

- The **Liquid Bulk Terminals** import straight run bitumen in large quantities and act as a distribution hub to other Terminals and directly to volume users. The bitumen will arrive at the Terminal by pipeline, ship or rail. A typical bitumen storage tank in this situation may be up to 10,000m3 capacity, flat bottom, cone roof, heated and insulated. Frequently they will have a double floor to prevent heat loss to the foundations.

- **End User** and **Distribution Depots** serve their local bitumen market or are major users of bitumen in their manufacturing process. Here the individual storage tanks tend to be a lot smaller, the total capacity usually depending on the resupply frequency from their wholesaler.

  The current trend is to use either narrow, vertical or horizontal tanks. For 'new builds' these types of tanks have several advantages;

  ❖ Manufacturers claim that a tank diameter of about 4m is the optimum as it offers the smallest possible surface area relative to tank volume;
  ❖ Tanks of this size are road transportable so are generally fabricated off-site in dedicated facilities, resulting in quicker build times;
  ❖ Newer tanks can have up to 300mm of insulation, weather-resistant cladding and efficient heating systems, meaning operating costs can be reduced;

  Horizontal tanks have traditionally been the ideal configuration. In general, horizontal tanks are straightforward to handle, install and move since they are permanently positioned on their skids. The normal maximum size of a horizontal tank is about 70m³

  Vertical tanks are space saving when space is at a premium. These small vertical tanks tend to be very high compared to their diameter. The use of vertical tanks sees a reduction in oxidation compared to horizontal tanks, due to the smaller exposed surface area. The normal maximum size of a modern (road transportable) vertical tank is about 150m³.

### *Heating*

Management is constantly targeting areas of high-energy consumption to reduce operating costs.

If bitumen is overheated, it can have a serious effect on the properties of the material, impacting its usability. Overheating can also lead to the build-up of deposits inside the storage tank and increased emissions. For this reason, bitumen should be stored at a temperature of at least 30°C below its flash point. The bitumen industry recommends storage between 130°C for the softest bitumen and 200°C for the hardest bitumen.

There are an endless variety of heating arrangements and permutations:

- **Direct heating**-the product is directly heated by combustion gases, flame radiation or electric heating elements, without any intermediary fluid. During the process, the combustion gases are expelled to the outside by means of a flue. When the product reaches the desired temperature, the burner is deactivated. Direct heating is the most efficient tank heating option. The heaters are immersed in the tank allowing for fast, even heat transfer.

- **Indirect heating**-an intermediate medium is used, which circulates in a controlled manner between the heater and the bitumen. This medium is usually a synthetic thermal (thermic) oil, designed to handle a wide range of temperatures without being degraded over time. Steam can be used but comes with a number of practical disadvantages.
  The main advantages of indirect heating are:
  - ❖ The convenience of locating the boiler;
  - ❖ Lower maintenance costs;
  - ❖ Higher efficiency;
  - ❖ Lower risk of overheating and elimination of local 'hot spots';

- **Steam**-super heated steam can be used to heat the bitumen. On Terminals taking deliveries by rail, there will always be a ready supply of steam available for wagon cleaning, so using it to heat the tanks makes sense. The steam circulates through heating coils laid out across the tank floor but leaks of steam release water condensation into the tank. Temperature control is course and consistent product temperatures are difficult to achieve. If a tank is coil heated, over time it is possible that coke will build up on the coils, making the coil less efficient and having a knock-on effect on energy consumption.

- **Thermic oil**-used as an intermediary between the 'boiler/heater' and the tank, thermic oil prevents corrosion on the inside of the heating pipes and allows more accurate control of temperatures.
  An oil fired or gas fired burner heats the thermic oil &it is circulated through the heating coil inside the bitumen tank by a hot oil pump. A single heater can easily heat several tanks as required in a continuous process. The temperature of the oil can be accurately controlled by an auto thermostat on the heater so the bitumen never gets over-heated. The oil is, however, quite expensive.

- **Electric**-claimed to be 99% efficient and this does not deteriorate over time. The electric heating elements are inside the sheath or the pipes, which are in direct contact with the liquid asphalt. In some cases (with narrow vertical tanks) the elements can be in a chamber below the tank. This avoids hot spots and allows defective heating elements to be changed without draining the bitumen from the tank.
  Electric heating systems can work on two or more levels: low-power for maintaining temperature, and booster power to raise the temperature if required. This multi-level heater configuration results in superior controllability and helps to minimise power consumption during operation.
  In the event of a tank going cold, suction heaters and backup electric immersion heaters may also be provided.

Most sites have one of the heating methods listed above or variations on them. On older sites, an alternative arrangement is for bitumen to be pumped to an external heater element, where it is heated directly or indirectly. For direct heating the bitumen passes through coils in the heater, and is heated by the combustion gases. For indirect heating, the heater circulates hot oil through the tube bundle of a shell and tube heat exchanger. The bitumen is in direct contact with the tube bundle and is heated before being returned to the tank.

This philosophy of taking the bitumen from the tank for heating has a couple of advantages. Heating coils in the bottom of the tank are omitted, as these are a common source of failure. The process of drawing out the bitumen and returning it to the tank via different pipe assists in the mixing of the bitumen within the tank. Shell and tube bundles have an additional advantage as the bundle can be 'pulled' from the shell and cleaned or repaired to restore thermal efficiency.

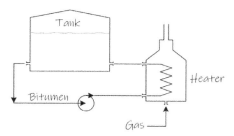

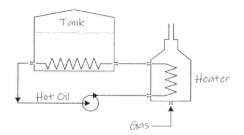

FIG.641    Bitumen heating using external gas or gas oil-fired heater

FIG.642    Bitumen heating using hot oil and in-tank heating coils

### Steam

If the Terminal operation requires steam for rail wagon unloading, a steam generator using hot oil as the heat source can be used instead of using a boiler. A steam generator system doesn't need a boiler or often a boiler operator. Eliminating the boiler and Operator will reduce OPEX.

### Mixing

Thermal stratification is an issue when tank temperatures have varied over time with inadequate mixing. Many tanks have test points on the shell, allowing samples to be taken at 3 or more levels of the tank, by Operators standing on the staircase.

Large tanks sometimes have side entry mixers. Vertical tanks usually have a centrally mounted, vertical mixer shaft with paddles at 2 or 3 points. Obviously, rotational speeds are slow because of the viscous nature of the bitumen, even at high temperatures. Horizontal tanks do not have mixers, relying on natural convection from the heater elements.

Another old-fashioned but reliable method of mixing is to have a 'bubbler' in the tank. Compressed air is blown into the tank or mixing vessel, and the passage of the air as it rises through the bitumen promotes the mixing of the tank contents.

### Metering and Pumps

Measuring liquid levels in bitumen tanks is notoriously difficult, as the hot vapour rapidly coats anything in the tank with bitumen, rendering some level systems inaccurate or ineffective. Makers of radar level transmitters claim good performance at high temperature and with a high vapour density in the air inside the tank that does not affect accuracy or stability.

However, many older locations still employ a simple float inside the tank and a marker board outside to show the liquid level, and are happy with the reliability and accuracy!

Pumps used for bitumen are frequently positive displacement types such as internal and external gear pumps. Where the bitumen is particularly viscous, pumps operate at lower r.p.m. Where a transfer pump is used to load road tankers, it will typically be rated at 30m³/hr. to ensure a tanker turnaround time of 60 minutes or less.

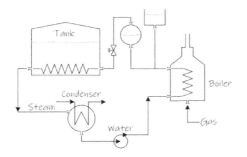

**FIG.643**   Steam heating using in-tank heating coils

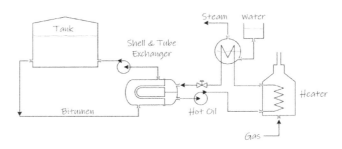

**FIG.644**   Hot oil system using the oil to generate steam for services and utilities

## Industry Specific Issues

### *Terms and Definitions*

Bitumen and asphalt are different things and frequently confused. Bitumen is a by-product of the crude oil refining process. In a distillation column, the crude oil is heated and the light ends evaporate and pass up the column. These vapours are turned back to liquids and are the basis of kerosene, gas oil, diesel and gasoline (via naphtha). The oil that stays at the bottom of the distillation column (known as residual oil) goes off to be used as feedstock on more exotic and expensive process plants. After going through these processes, and after any mineral oils have been recovered, what is left is bitumen (quite literally, the bottom of the barrel).

There are two materials known as asphalt. The most common is the material used for resurfacing roads, drives and parking areas. This is created by blending various combinations of bitumen, alumina, lime, stones, gravel and sand together. Bitumen is known for being strongly adhesive and resistant to damage from water and oil spills. This makes bitumen the ideal binder for asphalt.

Asphalt may also incorporate additives to improve performance (e.g. fibres, wax, polymers and pigments). Typically, asphalt is black, but it may also be coloured.

When constructing roads, different grades of asphalt are used for the road base, the binder and the surface courses.

However, what is called "asphalt" can also be found as a naturally occurring material in some lakes or extracted from limestone and sandstone. In practice, it mostly comprises bitumen that has seeped up through the ground. In some countries, this may be the major source of bitumen for road making.

Bitumen can also be confused with tar. The principal difference for the casual observer is that bitumen can be a solid form or viscous but tar is always viscous.

## Measurement and Standards

Bitumen is usually defined by its;

- Viscosity;
- Density;
- Penetration;

Bitumen penetration is measured using a 100g needle dropped onto a sample for 5 seconds. Lower 'pen' (penetration)bitumen is, therefore, stiffer and has a higher softening temperature.

The most common specifications for bitumen are **ASTM D946-09** and **EN 12591**.

Under EN 12591, a pen of 100-150 bitumen is the preferred grade for most road-building applications. Pen 40-60 is used for heavily trafficked roads, and pen 160-220 is often used for hand lay work as it is easier to apply.

Bitumen of different grades can be mixed and blended to produce intermediate pen grades as required by the Customer. The most common arrangement is for the Terminal to receive just 2 grades of bitumen, and then blend in batches to meet demand as orders are received. For instance, if the Terminal receives shipments of 40/60 and 160/220 bitumen, intermediate grades are straightforward to blend.

Testing on the Terminal is simple and quick so there is no reliance on external laboratories.

## Safety

Transferring, handling, and storing hot bitumen exposes Operators and Hauliers to inherent hazards:

- Thermal burns;
- Froth-overs and boil-overs due to contamination by water;
- Formation and release of hydrogen sulphide/hydrocarbon fumes;
- Formation of flammable vapours;
- Creation of ignition sources (see below);
- Falls from elevated work areas such as the tops of tankers;

Pyrophoric deposits are formed on the roof and walls of bitumen storage tanks due to the condensation of bitumen vapors and the reaction with iron. These can bea potential ignition source.

If water has collected in a pipe or vessel and bitumen is introduced, the water flashes off to steam, causing seals and gaskets to fail and pressure relief valves to discharge a shower of hot bitumen. Due to the viscous nature of the bitumen, it is impossible to quickly remove it from the skin or lower the temperature, so even small burns can be quite severe.

There is little reason for Operators to be anywhere near the bund or tanks during normal operations so the actual risk in the tank farm is low. However, loading trucks with bitumen requires special vigilance and precautions.

## Value Added Services

**Foamed bitumen** has proven an ideal method to improve the handling qualities of low temperature road making asphalt. Prior to the bitumen being discharged into the mixer with the aggregate, the foaming process requires water to be added to the bitumen in accurately metered quantities.

Alternatively, the foamed bitumen can be produced using compressed air blowing through the bitumen at high temperatures and in this case it is marketed as 'oxidised' bitumen.

The resulting asphalt flows easier and creates a flexible, penetrating base layer.

There are some instances where it is beneficial to transport bitumen in a **solid form**. As bitumen can set without harming its physical and chemical properties, this is usefull when storing and delivering small quantities where continuous heating to keep it in liquid form is not practical. Obvious uses are the production of 'easy melt' bitumen blocks that weigh about 10kgs. so are easy to manhandle. These are used by road construction crews, road repairers and roofing contractors where the brick is melted down in a steel 'pot' heated with propane. Blocks (including foamed or oxidized) are sold individually or by the pallet.

Along the same lines, some Terminals may prepare cargos of Bitumen for shipment in drums in solid form. When bitumen is delivered in 200 litre barrels, to remove it from the barrels it is necessary to heat it. To ensure there is no deterioration of the quality of the bitumen during reheating, the heating process is achieved using circulating hot air in an enclosed unit.

### Seasonality

Most Bitumen Terminals will only supply to manufacturers and construction companies in bulk. By bulk, I mean the lorry load or rail tanker load. Road repairs and new road works are carried out mostly in the summer, meaning that throughput in the winter months can be quite low.

Customers take the hot liquid bitumen in bulk for processing at their yard. Road construction companies will blend it with aggregates for road bases and wear coats, or decant the hot liquid into smaller vessels for use at the road works (with gas burners) to prime existing surfaces. Roofing companies will take the bitumen direct by road to the site where it can be laid and leveled by hand. Specialist companies who make playgrounds, outdoor tennis courts, race tracks and multi-use play areas may add colouring or other additives before use. These companies can produce surfaces that have smooth, porous, low texture, low friction finishes by using small aggregate in a dense asphalt matrix.

## Terminal Layout

### Liquid Bulk Terminal

The economic model is to import in bulk, store, break the cargo into smaller parcels and distribute to end users or smaller bitumen Terminals serving regions or districts. In some cases they will add value to the bitumen by additional processing to met Client requirements or packaging for export.

Imports will generally arrive at the Terminal via pipeline, ship or railway wagon. The exported product, either straight run bitumen or bitumen blends, will leave the Terminal by road tanker or rail tanker. Therefore, in most cases the Terminal layout will be dictated by:

- Its proximity to a port or harbour or;
- Its proximity to a refinery or;
- Road tanker access or;
- Rail tanker access;

As the point of the Terminal is to buy and store product in bulk, tanks will be quite large (typically up to 10,000m³). The number of tanks can vary depending on the market it supplies, but many Bitumen Terminals will have just two tanks, one for the low pen product and one for the high pen product. Intermediate products are produced by blending high and low pen products together. Although there are 'standard' grades, Customers can specify the specific grade that suits their subsequent processing.

The equipment for loading road and rail tankers has been considered in detail elsewhere but bitumen handling operations are far simpler.

Rail wagons and road tankers used to deliver bitumen are insulated and have heaters to maintain product temperatures during product movement.

For **offloading**, bottom loading arms or hoses are used and the product is pumped away to the storage tank after its quality has been verified. There is usually some degree of hangup on the walls of the tank, especially the rail wagon which has probably taken a number of days to arrive at its unloading point. Standard practice is to return the tanks empty and clean, ready to accept another cargo, so the rail wagons are generally steamed during unloading. This raises the temperature of the tank, allowing gravity to complete the tank emptying. After unloading, the transfer lines (which form a ring) are blown down with air to stop them plugging.

For **loading**, top loading arms and hoses are used, usually counterbalanced to make their handling easier. Unlike ground fuels, there is no blending required at the loading rack and no accurate metering of quantities or compartmentalization of cargos. As we are dealing with a low-cost, high-volume cargo, automation of the systems is seldom justified, and most operations are manual.

Bitumen Terminals are not good neighbours and are usually sited away from residential areas to avoid complaints about the smell. Partly as a precaution and to minimize any potential GHG emissions, the ends of the top loading arms are frequently in the form of a tapered plug to fit the tank's circular hatch. These plugs have vapour extraction lines to extract fumes as the tanks are being filled.

To the casual observer the loading arms and loading bays usually look just like those used for gasoline and diesel, except dirtier!

### End User and Distribution Depots

Both of these locations will carry smaller volumes of bitumen and will most likely have narrow, vertical tanks or horizontal tanks.

**End users** may use the bitumen as part of their manufacturing process (such as an asphalt paint supplier, bitumen sealer or making the backing for carpet tiles). In this case they will have enough tanks to allow uninterrupted operations between deliveries from their supplier at the Liquid Bulk Terminal.

Other end users (the majority) will use the bitumen as a binder when mixed with aggregate and sand. In this case, the Terminal will have hoppers for storage of the dry bulk materials as well as mixing vessels where the components are brought together. At the specialist end of the market, some companies produce high-tech sports surfaces and coloured asphalt.

**Distribution Depots** will in all senses be mini versions of the Liquid Bulk Terminals. They may import cargos of straight run bitumen or blended bitumen by road or rail, and will export their cargos by road. Their Customers will be individual road work Contractors or roofing Contractors who will pick up a cargo from the Depot to be taken direct to the job site.

FIG.641　A horizontal Bitumen tank with direct fired heater being used during road construction

FIG.642　A modern Bitumen Terminal using vertical tanks to reduce the footprint
Photo; Ammann

### *Firefighting*

Firefighting facilities can be limited as bitumen does not catch fire easily. In the event of the bitumen in a tank catching fire, external cooling sprays are of no benefit with the insulated and clad tank.

### *Secondary containment*

The main source of leaks are the flanges of interconnecting pipework and the seals of rotating equipment. As a consequence, spills tend to be localized. In the event of a leak into the secondary containment, the bitumen quickly solidifies and does not 'flow' and the floor of the secondary containment does not need to be watertight.

An interesting choice seen at one Terminal was the use of lightweight aerated lava rock or pumice. The lava rock had an open texture with a large surface area. When bitumen is spilled in the bund, it quickly solidifies. To clean it up, the solidified bitumen and the lightweight rock is shovelled out and replaced (at low cost) by fresh material.

## 6.5    Edible Products

**Tank Design**

### Compliant industries

An **edible item** is any item that is safe for humans to eat or drink.

"Edible" does not indicate how an item tastes, only whether it is fit to be eaten. Typical categories of liquid or semi-liquid bulk storage include:

- Food including processed food;
- Condiments;
- Beverages including fruit drinks;
- Diary;
- Vegetable Oils;
- Brewing;
- Wine;
- Confectionary;

The common element of these types of tanks is that they all have to meet the standards for Food Contact Material (**FCM**s). In addition, they have to be designed to minimize the risk of spillage and product contamination.

In addition, there are industries such as cosmetics and pharmaceuticals where similar rules apply for the cleanliness and sterilisation of contact surfaces. Similar rules may also apply to some types of animal feed where there is a risk of microbial contamination.

### Definitions

Whilst some edible products are stored in bulk in Terminals (such as sunflower, corn, rapeseed, olive and coconut oils), the majority of tanks in the food industry are used as components in the manufacturing process.

A generalization is that;

- **Tanks** are used to store liquids and semi-liquids in bulk before processing and after processing, and as buffer storage between stages of the manufacturing process;

- **Hoppers** and silos are used to store dry products and materials in bulk. They can look similar to tanks, but have cone-down bottoms and are generally elevated to allow gravity feed of their contents into the manufacturing process;

- **Vessels** are used to process the components into a finished product. Vessels can have stirrers or agitators and be used for heating or cooling mixtures;

- All the above are connected by pipes and pumps, which must meet the same FCM standards as the tanks, hoppers, silos and vessels;

- In addition, there are ovens, conveyors and packaging lines that we need not concern ourselves with;.

There is little point in considering the types of tanks or equipment layouts as they will vary from one location to another. Instead we will consider the selection of materials and common design details to prevent microbial contamination that will be common across the entire industry.

## Material standards

During contact, molecules from materials can migrate into foods or beverages. FCMs should be sufficiently inert so that their constituents neither adversely affect consumer health nor influence the quality of the food. Because of this, most countries have food safety regulatory agencies that monitor materials for food safety.

In the **United States**, the most recognized and frequently encountered food safety agency is the Food and Drug Administration (**FDA**). The FDA determines the appropriate use of materials for potable beverage and foodstuff processing, handling and packaging. The FDA does not provide government inspection and certification of materials used for food contact applications. Instead, the agency sets rules and guidelines regarding appropriate material composition, properties and uses.

The National Sanitation Foundation (**NSF**) certification requires testing to NSF standards for contact with drinking water, food and so on. NSF certification is provided by NSF accredited third-part certification organizations such as the Water Quality Association.

**NSF/ANSI-51**, Food Equipment Materials establishes the minimum health and sanitation requirements for materials used in the making of commercial food equipment. Materials must not contaminate food nor make food equipment difficult to clean and sanitize.

In the USA you may also encounter regulation by the United States Department of Agriculture (**USDA**).

In **Europe** the general requirements for all food contact materials are laid down in EU Framework Regulation EC 1935/2004, including plastics used as packaging materials. The safety of FCM is evaluated by the European Food Safety Authority (EFSA). The UK uses EU food safety standards.

For the manufacture of tanks, hoppers and vessels, stainless steels meeting **SAE 304** and **316** are almost universally used because;

- Resists acidic food;
- No rust or corrosion;
- Does not flavour the food;
- Withstands high temperatures for sterilisation and cooking;
- Withstands low temperatures for frozen dairy products;

Where the ability to withstand extreme temperatures is not required, potable water tanks with NSF approval can be made from a variety of materials such as:

- Glass fused to steel (bolted);
- Epoxy coated steel (bolted);
- GRP (bolted);
- Polyethylene;
- HDPE;

## Design considerations

There are three potential hazards to consider with the design of equipment:

- **Chemical**-can be caused by cleaning chemicals or from lubricating fluids. Avoided by staff training and operating instructions;

- **Physical**-principally glass, insects, pests, metal and dust, Avoided by a sealed manufacturing process and environment;

- **Biological**-at harvest or slaughter, products can have a variety of microbes in or on them. Some will cause eventual spoilage and some can cause sickness to consumers. In some cases, there will be design flaws that will cause contamination to be carried over from one batch to another, whilst microbes can be transferred by workers due to incorrect operating procedures.

  - ❖ **Draining**-equipment must be designed to be self-draining to make it possible to remove all residues of products and chemicals. For tanks, this means they should be a self-draining cone down design and lapped floor plates replaced by butt welded plates. If pipework cannot be laid to a fall, additional drain points must be provided or different operating/cleaning procedures adopted.

  - ❖ **Surface roughness**-to be cleaned without difficulty, surfaces must be smooth and free from crevices, sharp corners, protrusions, and shadow zones. Generally, the cleaning time required increases with surface roughness. Porous surfaces are unacceptable. When surfaces are not clean, microorganisms may be protected from destruction by heat or chemicals;

  - ❖ **Crevices**- cannot be cleaned, and may retain product residues. Metal to metal joints leave narrow but deep crevices, so food-safe PTFE gaskets should be used;

  - ❖ **Screw threads**- the use of screw threads and bolts in contact with the product should be eliminated;

  - ❖ **Sharp corners**-should be avoided. Smooth rounded corners are easier to clean;

  - ❖ **Flat horizontal surfaces**-should be avoided and always sloped to allow liquids to drain away from hatches and man doors;

  - ❖ **Dead areas**-are difficult to clean and should be eliminated where possible;

  - ❖ **Bearings and sliding surfaces**-within a vessel or tank should be avoided or special cleaning instructions developed. If a bearing in contact with the product is unavoidable, it should have food safe lubricants. Preferably, magnetic drives should be used to reduce the number of mechanical seals;

## 6.6    Refineries

### A Different Perspective

In the case of Refineries, we can stop talking about Terminals, and instead refer to them as **Tank Farms**. The Refinery may contain many of the same things we have already considered i.e. bitumen storage, road and rail loading facilities, pipelines, and import/export by ship. In practice, however, none of this counts.

At a Refinery the Tank Farm has only one purpose-it is there to facilitate the work of the process units. There are many different types of process units, as well as many variations, proprietary techniques and hybrid models. They convert low-cost 'crude' products into high-value refined products. Many sorts of refineries exist, producing many different 'refined' products such as sugars, palm oil and specialist chemicals. To simplify things, we can again ignore all of this.

The principal purposes of the Tank Farm are:

- To hold unrefined materials waiting to be processed. To be a buffer between receipt and processing;

- To hold the refined product. To be a buffer between production and export;

So why the heading '**a different perspective**'?

### The Refinery Perspective

The Engineers and Operators at Refineries have only one interest and it isn't the Tank Farm.

It is the process units. They are at the centre of the Refinery and it is how the Refinery makes its money. Tank Farms, common services and all the other essential plant and equipment don't even get a look in.

Most Refineries recruit young inexperienced Process (Chemical or Petroleum) Engineers straight from University. They are usually sent to the Tank Farm (or offsites as it is frequently known in the petrochemical industry). As a result, the Tank Farm is regarded as the nursery, the obligatory punishment, a period in Purgatory. A place the young Engineer can make his inevitable mistakes without it costing anyone too much money before the Engineer is moved on to a 'real' job. The young Engineer's primary concern is to get away from the Tank Farm as quickly as possible, to do more interesting and rewarding work on the 'process' units.

For Process Engineers, the principal document in any design is the Process Flow Diagram (PFD). The PFD has an arrow going into the box that represents the tank and an arrow coming out. Nothing happens within the box. The Process Engineer is also concerned with mass flow, changes in pressure and changes in temperature. Again, nothing happens within the box. What goes in is what comes out. All very uninteresting for a novice Process Engineer.

Process Engineers get promoted to process units and ultimately promoted to management. So in most Refineries the senior management is probably full of people who have no technical interest in the Tank Farm, other than when a fault exists that may impact their beloved process units.

Most Operators start in the Tank Farm. The clever, bright and hardworking may end up being promoted to work in a Process Unit, but promotion is slow. Most 'Chief Operators' can be on a particular process unit for 15-20 years and there is only one per shift.

There is an unspoken but very real hierarchy within most petrochemical Refineries. The 'cracker' Operator is at the top of society. The 'distiller' Operator is halfway down the pecking order. The 'tank farm' Operator is very much at the bottom of the pile.

Many Refineries have different Control Rooms for the Process Units and the Tank Farm. In some, there is no movement of staff between the two and being posted to the offsites is very much like being in the leper colony. Some sites claim to rotate staff between Control Rooms, although I have my doubts this actually happens in practice.

So it is clearly established for Engineers and Operators all through their career progression that dealing with tanks is a menial role, not something to aspire to. And because the offsites usually include the marine facilities, product transfer, common services and infrastructure, all the most interesting parts of a Terminal are lumped together into a dismissive parcel.

FIG.661    The Operators of the process units at Refineries always look down on the Operators working in the tank farms and offsites
Photo; Koole

## The Terminal Perspective

The Terminal Owner has a completely different perspective.

Where the Refinery Operators see no technical challenges, the Terminal Owner and his staff see endless problems and complications, difficult technical issues to be resolved under constant time and cost pressures.

Why the difference? Well, it's the approach to tanks.

Ask a Refinery Operator and he will tell you tanks are a necessary evil. But he doesn't have enough tanks. He will never have enough tanks. He doesn't care whether they are empty or full, he just needs more. How many is enough? As many as necessary to make sure the process unit always runs at maximum capacity, unconstrained by limited raw stock or storage for refined product.

Got a faulty blend or off spec product? Just stick it in a tank and leave it for the time being. What if all the tanks are full of off spec product and bad blends? Or the shipment of that 'special' export blend is delayed? Well, clearly the problem is the Refinery's Tank Farm doesn't have enough tanks.

So the Refinery Operator wants lots of tanks and is quite happy if they stay empty most of the time, but the Terminal Operator sees a tank that is empty and worries it is not earning its keep.

The Terminal Owner needs to have customers for every one of the tanks. Tanks need to be cleaned, inspected and put back into service as quickly as possible so they can earn money. Tanks need to meet the Client's requirements. If they don't then new tanks that do meet Client requirements must be designed and built as replacements. New Customers need to be found and old customers need to be kept happy. Product movements need to be well organized, consistent and quick. Everything needs to be done with the minimum of manpower, at minimum cost.

FIG.662    No matter how many tanks there are in a Refinery, it is never considered enough
Photo; Koole

The Terminal has the same business overheads and the same workload as a Refinery but without the 'cash cow' of a process unit to generate the gold. In short, the Terminal Owners have all the work and risk but a fraction of the profit of a Refinery.

At the same time, Refineries tend to get away lightly during inspections and audits. Everyone concentrates on the 'high risk' process units and Tank Farms barely get noticed. By comparison Terminals don't have anything to divert the Inspector's attention and will be picked up on small deficiencies most Refineries will get away with.

Now don't get me wrong. By background, I am a Refinery man. If picking a team for design or maintenance, I would ALWAYS pick people with a Refinery and Oil & Gas background. But the skills and competencies of the Terminal Operators and their Maintenance colleagues are frequently overlooked and undervalued.

Newly qualified Engineers, hoping to get a better understanding of the bulk liquid storage market could do a lot worse than ask a Terminal man, then listen closely to their answer.

•••••••●●●●●●●●●•••••••

# APPENDICES

# APPENDIX A

# *Physical Properties*

## Units of Measure

For convenience, we have used 'International System' (S.I.) units throughout. Tanks sizes are referred to by their volume i.e. 10,000m³. For convenience **Appendix B** contains typical conversion factors.

## Density and Specific Gravity

The density of a liquid is its mass per unit volume. Water is accepted to have a density of 1g/cm3 (or 1g/mL or 1000kg/m3) at 40C. The exact density can vary slightly on temperature and air pressure, but for practical purposes, this can be ignored.

The specific gravity of a liquid (sg) is defined as the ratio of the density of a liquid compared to the density of water. As water has a density of 1.0 g/cc, the specific gravity of a substance that is 1.5 times heavier than water would be equal to 1.5. The specific gravity of a substance that is 1.5 times heavier than water would be equal to 1.5.

Specific gravity has no units because it is the ratio of two density numbers.

The density of a liquid is important in the design of tanks as liquids with greater density require thicker tank shells. If a tank operator changes the liquid in a tank for a denser liquid, the maximum liquid level in the tank has to be lowered, to reduce the hydrostatic pressure acting on the tank shell.

In the petroleum industry, a common indicator of specific gravities is known as API gravity. This is used for measuring the relative density of petroleum liquids, expressed in degrees so that most values would fall between 100 and 700API gravity. Using this scale water has an API gravity of 10.

The API gravity is most commonly used to identify the properties of different types of crude oil. The arbitrary formula used to obtain this is;

*API gravity = (141.5/SG)-131.5 where SG is the specific gravity of the fluid.*

## Temperature

Both Celsius and Fahrenheit scales are in common use although for consistency we will keep with Celsius. The Celsius scale defines 0°C as the freezing point (triple point) of water and 100°C as boiling point. Fahrenheit arbitrarily assigns freezing of water to 32°F and boiling point of water to 212°F.

It is sometimes necessary to convert Celsius to an absolute temperature scale called *Kelvin*. The Celsius scale reads 0°C while the Kelvin scale reads 273°K. The equivalent 'absolute' scale for Fahrenheit is Rankine.

Tanks are designed to store liquids over a wide range of temperatures. Liquefied Natural Gas (LNG) is stored at -162⁰C whilst Asphalt is stored and blended at 130⁰C to 160⁰C. Tanks usually store products at ambient temperatures, but even these can vary widely depending on location, season, weather and time of day.

At low temperatures, material selection becomes an important design consideration to avoid brittleness. At high temperatures, corrosion is accelerated and thermal expansion must be taken into account.

## Vapour Pressure

**Vapour pressure** is a measure of the tendency of a material to change into the gaseous or vapour state, and it increases with temperature. A substance with a high vapour pressure at normal temperatures is often referred to as volatile.

Vapour pressure is important in the design of storage tanks as it can affect the selection and design of the tank and roof and evaporation losses. For instance, the vapour pressure can be critical in deciding whether a floating roof is required. For flammable liquids, vapour pressure is important in classifying liquids to determine the fire hazard they pose.

The boiling point of the liquid is also critical. If the vapour pressure exceeds the gas pressure above the liquid, then vapour bubbles appear in the liquid and the liquid 'boils'. It is important therefore to always express the vapour pressure at a particular temperature.

Vapour pressure is important for gasoline-powered, especially carbureted, vehicles. High levels of vaporization are required for winter starting and but lower levels are desirable in summer to avoid 'vapour lock'. Refineries adjust the vapour pressure of gasoline seasonally to maintain gasoline engine reliability.

There are two other types of vapour pressure that the reader may come across: **True Vapour Pressure** (TVP) is a common measure of the volatility of petroleum distillate fuels. It is determined by ASTM D 2879. **Reid Vapour Pressure** (RVP) is similar but it excludes dissolved gases. RVP can vary significantly from TVP and is determined by ASTM D 323.

Being based on American standards, TVP and RVP are commonly reported in pounds per square inch absolute at 100⁰F but kilopascals(kPa) may also be used.

Vapour pressure has become more significant as environmental controls have tightened. Products with higher vapour pressures result in an increase in emissions and some countries have maximum vapour pressures for which specific tank designs are permitted.

## Absolute and Gauge Pressure

Pressure can be expressed as an absolute or relative value.

Although atmospheric pressure (also known as barometric pressure) constantly fluctuates, a standard value of 101,325 Pa (or more commonly 1 atm) has been adopted at sea level (14.7 psia in U.S. units). This is an absolute figure, meaning it is above a complete vacuum. Most ordinary measuring devices do not measure the true pressure, but the pressure relative to atmospheric pressure. This relative pressure is called *gauge pressure*.

Vacuum is a relative pressure as well. For tank work millimeters of water column are commonly used to express the value of pressure or vacuum in the vapour space of a tank because the pressures are usually very low relative to atmospheric pressure.

The difference between the pressure inside a tank and atmospheric pressure is called *internal pressure*. When the internal pressure of the tank is negative, it is simply called a *vacuum*. The pressure is measured at the top of the liquid to exclude the hydrostatic head which increases at the bottom of the tank.

As tanks can be very large structures, even small pressure increases can exert large forces on the structure. This is one of the reasons venting of atmospheric tanks can be important. Even a small vacuum inside the tank can result in significant additional external loads on the roof and shell causing the tank's buckling and eventual collapse.

Where the vapour space of a tank is open to the atmosphere or freely vented, the internal pressure of the tank is always 0 barg or atmospheric. However, most tanks are not open to the atmosphere but are fitted with a device sometimes known as a **pressure-vacuum valve** or a **breather valve**. The purpose of this valve is to ensure the tank is not subjected to excessive internal pressure or excessive vacuum. In the event of a tank over pressurization (due to fire or explosion), the joint between the roof and the shell plates is frequently designed to fail. This reduces the likelihood of the shell/bottom plate joint failing, which would result in a catastrophic loss of the primary containment. Cone roof tanks designed to fail at the shell/roof joint are described as 'frangible'.

The maximum tank internal pressure or vacuum is built into the tank design code. Although requirements change from country to country, a rough rule of thumb is that any storage vessel with an internal pressure greater than 1 atmg or 15 psig is generally classified as a pressure vessel and has to be designed, built and operated under specific pressure vessel regulations.

Section VIII of the ASME Boiler and Pressure Vessel Code is a typical example.

## Flash Point

A liquid does not burn. It is the vapour that mixes with oxygen in the atmosphere above the liquid that burns. As a liquid is heated, its vapour pressure and consequently its evaporation rate increase. The minimum temperature at which there is sufficient vapour to allow ignition of the air/vapour mix near the surface of the liquid is called the **Flash Point**.

For flammable and combustible liquids, the Flash Point is the primary basis for classifying the hazardousness of a liquid. Low flash point liquids are high fire hazard liquids.

## Classification of Products

Different international standards identify dangerous products in different ways. Product classification can vary from country to country and some international oil companies have their own classifications:

i)  One classification frequently used by oil companies is **NFPA**, where the different categories are used to identify the specific fire risks associated with the storage of each product. The categories are based primarily on the flash point of the liquid i.e. the minimum temperature at which sufficient vapour is given off to form an explosive mixture with air.

**Flammable Liquids**

| | |
|---|---|
| Class 1A | Flash point <73°F (22.8°C) |
| | Boiling point <100°F (37.8°C) |
| Class 1B | Flash point <73°F (22.8°C) |
| | Boiling point >100°F (37.8°C) |
| Class 1C | Flash point >73°F (22.8°C) but <100°F (37.8°C) |

**Combustible Liquids**

| | |
|---|---|
| Class II | Flash point >1000F (37.80C) but <1400F (600C) |
| Class IIIA | Flash point >1400F (600C) but <2000F (930C) |
| Class IIIB | Flash point >2000F (930C) |

ii)     When transporting products by road or rail, many countries mandate the use of warning signs to comply with the **"Transport of Dangerous Goods"** (TDG) Model Recommendations. You will frequently see road tankers marked with the following classifications;

| | |
|---|---|
| Class 1 | Explosive |
| Class 2 | Gases |
| Class 3 | Flammable Liquids |
| Class 4 | Flammable solids |
| Class 5 | Oxidising substances |
| Class 8 | Corrosive substances |
| Class 9 | Miscellaneous |

These warning signs also carry a globally harmonised system of classification and labeling of chemicals. This provides additional information on the hazard including, where appropriate, the degree of the risk.

In the EU this is governed by Regulation (EC) 1272/2008

••••••••••••••••••••••

# APPENDIX B

# *Conversion Factors*

**Length**

| | |
|---|---|
| 1 inch (in.) = | 2.54 cm |
| 1 foot (ft.) = | 30.48 cm |
| 1 yard = | 914.40 cm |
| 1 cm = | 0.3937 in. |
| 1 metre = | 39.37 in. |

**Volume**

| | | | |
|---|---|---|---|
| 1 litre (l) = | 1000 cm$^3$ = | 61.02 in$^3$ = | 0.03532 ft$^3$ |
| 1 cubic metre (m$^3$) = | 1000 l = | 35.32 ft$^3$ = | 6.289 barrel (US) = |
| 1 cubic foot (ft$^3$) = | 7.481 US.gal = | 0.02832 m$^3$ = | 28.32 l |
| 1 US gal = | 231 in$^3$ = | 3.785 l | |
| 1 British gal = | 1.201 US gal = | 277.4 in$^3$ | |
| 1 barrel (US) = | 0.15898 m$^3$ = | 42 US Gal | |
| 1 barrel (Imperial) = | 0.1636 m$^3$ = | 36 British gal | |

**Mass**

| | |
|---|---|
| 1 kg = | 2.2046 pounds (lb.) |
| 1 lb. = | 453.6 gm. |

**Force**

| | |
|---|---|
| 1 US short ton = | 2000 lb. |
| 1 US long ton = | 2240 lb. |
| 1 metric tonne = | 2205 lb. |

**Power**

| | |
|---|---|
| 1 horsepower (HP) = | 745.7 watts |
| 1 kilowatt (kw) = | 1.341 HP |

## Pressure

| | | | |
|---|---|---|---|
| 1 lb./in²= | 5.171 cm mercury = | 27.68 in water= | 6.894 kappa |
| 1 lb./in²= | 0.0689 bar | | |
| 1 bar= | 14.5 lb./in² | | |
| 1 atmosphere = | 14.70 lb./in2= | 76 cm mercury = | 406.8 in. water |
| 1 kPa= | 0.75 cm mercury= | 0.145lb./in²⁻ | 0.01bar |

## Specific Gravity (examples)

| | |
|---|---|
| Ether | 0.71 |
| Gasoline | 0.75 |
| Isopropyl alcohol | 0.78 |
| Acetone | 0.79 |
| Kerosene | 0.8 |
| Alcohol | 0.81 |
| Diesel | 0.86 |
| Olive Oil | 0.91 |
| Peanut Oil | 0.92 |
| Cod Liver Oil | 0.92 |
| Castor Oil | 0.96 |
| Water (fresh) | 1 |
| Water (sea) | 1.03 |
| Glycol | 1.03 |
| Phenol | 1.07 |
| Glycerin | 1.25 |
| Hydrochloric Acid | 1.37 |
| Nitric Acid | 1.42 |

••••••••••••••••••••

# APPENDIX C

# *Auditing Existing Terminals*

## C.1    Preface

This section may seem out of place with the remainder of this book.

However, if you work at a terminal you will sooner or later be working with an Auditor (ISO 9000 audits tend to drag in the Terminal's whole management team). And if you work at an Engineering Consultancy you may well be asked to carry out an audit of a liquid storage Terminal as part of a Client's 'due diligence' or 'technical review'. These activities are explained in **Sections C.2** and **Section C.3** below.

Engineering Consultancies look down on the bulk liquid storage industry. If a Terminal audit comes up, it will in most cases be given to the more junior employee as a training exercise or as career development. Older and more experienced employees of the big Engineering Consultancy tend to consider 'terminal' related assignments unworthy of their talents.

Thesenotes areintended to provide guidance to potential Auditors on how to organise a technical review together with some relevant background information. The description is neither complete nor detailed but included within this text to be indicative of the process.

For those Engineers working in the bulk liquid storage industry, it may be instructive to understand how the Auditors work and what they are looking for.

At some terminals, visits by external auditors are a frequent event. At other terminals, the presence of auditors is unusual and resident staff can feel defensive. This is seldom warranted as any experienced professional auditor will not be judgmental. In most cases, they have been subjected to audits themselves and are unlikely to aggressively seek out wrongdoing or point the finger.

There are many sorts of audits carried out on Terminals, but we are only interested in two.

- '**Due Diligence**' audits are carried out on behalf of Investors and Shareholders. See **Section C.2**;
- 'Technical' audits are carried out on behalf of the Terminal Owner. See **Section C.3**;

## C.2    'Due Diligence' Audits

Bulk liquid storage terminals have been attractive investment vehicles in the last 20 or 30 years for insurance companies and pension funds as they have shown continual growth. However, it is not unusual for Investors to move on every 7 years or so. Long-term investment in companies is not the current trend.

A Terminal Owner could be seeking funds or investment to improve or extend their production facilities. Alternatively, the Terminal could be the subject of merger and acquisition activity.

157

The Terminal may need **Lenders** or **Investors** because:

- They need to extend production to meet current demand;
- They have identified a new potential market or customer;
- They need to upgrade or refurbish to meet increased legislation;
- They need to get efficiencies to remain competitive;
- An existing shareholder wants to liquidate their assets;
- The Terminal Owner plans to acquire a competitor;

It is normal for potential Investors or Lenders to carry out a 'Due Diligence' review into the viability of the Terminal. This will consist of a **financial audit** they do themselves but they will frequently employ an Engineering Consultancy to perform an audit of the **Terminal's management and performance** as well as the **condition of assets**. The Consultancy may be employed by the potential Investors or Lenders, or directly by the Terminal Owner.

If the new funds are for the improvement or extension of their existing facilities, the auditmay well include an assessment of the feasibility, cost, and schedule of the new installation. New and planned Terminals are considered in **Appendix C-New Terminals**.

For the technical auditors, the current business operations give a good guide to how well the business is run, based principally on current and past performance and the condition of the assets. The key objectives will be:

- To prepare a report, aimed at a non-technical audience;
- To put potential risks into context;
- To identify avoiding actions where feasible;
- Suggest mitigating actions where risks cannot be avoided;

These audits tend to be carried out by groups of Engineers who may not have an in-depth understanding of how the specific business operates. It is in these cases that misunderstandings can arise. The site staff should appreciate that in some circumstances they will be helping to train the Auditors about their site and silly questions by the Auditors are inevitable!

A lot of information is gathered by the Auditor from the documentation submitted by the Terminal. However, site visits are one of the best methods for the Auditor to get a real appreciation of the business, to meet the local Management Team for one-to-one briefings and the only way to understand the condition of the asset.

## Business Background

When a Terminal faces competition, it must cut costs and innovate. One solution is to acquire competitors so that they are no longer a threat. This can also bring benefits by economies of scale, lower overheads or increased market share.

The term Mergers and Acquisitions (M&A) is a general term that describes the consolidation of companies or assets through various types of financial transactions including mergers, acquisitions, consolidation, purchase of assets, and management acquisitions.

The Client may appoint an investment banker as an M&A Adviser. The M&A Advisor may prepare an Investment Memorandum to attract external investors, either specifically targeting a known group or soliciting willing investors. It is important to recognise that an IM is part of the sales literature. Where negative facts are obvious and inescapable, the M&A Advisor is skilled in presenting the information in the most positive way possible.

In addition, the M&A Advisor can set up and manage the Vendor's Data Room (VDR). An increasingly popular trend is for the VDR to contain minimal documentation to make the Investor's Due Diligence more difficult. In some cases, the available documentation has sections redacted, and key documents are missing or incomplete. In extreme cases, the M&A Advisor will 'pad' the VDR with documentation that is irrelevant, out of date or contradictory, and then fail to respond to requests for clarification.

The documents most likely to be missing are those that provide an overview or allow understanding of the other documents provided. A typical example of this would be to provide a large but incomplete set of old inspection reports without providing an up-to-date list of assets.

To compound this problem, it is not uncommon for the number of questions to be limited, with arbitrary deadlines imposed. This means that even if the Auditor can identify missing information and request specific documents, it is unlikely they will receive anything.

Whilst we all accept the logic of "Caveat Emptor", it is a puzzle why potential Investors should tolerate restricted access to the Vendor's documentation.

Under the influence and guidance of the M&A Adviser, site visits, where allowed, are constrained, of limited duration and shared with other Investors. Where site visits are permitted it is not uncommon to find that no dialogue is possible with the site's operational or maintenance staff. Meetings are restricted to senior management who have little or no knowledge of the detailed history or performance of the asset. During these site visits the opportunity to see the assets is frequently limited to a short bus ride around the site, the Vendor using "safety", time or logistics as the excuse to justify the restrictions.

To the Auditor, it appears the sole objective of the M&A Advisor is to make their life more difficult. Whilst this may not be their primary objective, it is frequently an unhappy consequence.

## Role of the Auditor

Where the Audit is being carried out as part of a 'Due Diligence' review, the Auditor is usually employed by an Investor or a Lender, who may be either a **Seller** or a **Buyer**.

- A **Seller** will expect their Auditor to describe the asset positively:

- A **Buyer** will expect their Auditor to highlight negative issues:

These contrasting philosophies should guide the Engineer through the audit, as it will affect their approach and the tone of the final report.

The Auditor's team will prepare a report for the Employer.

If the Employer is a 'Seller', the report will be used to 'sell' the investment opportunity to interested parties. As such, the report by the Auditor forms part of the sales literature of the Seller.

If the Employer is a 'Buyer', the report will help negotiate the price down or to identify costs and technical problems not previously identified.

## *Whose Side is the Auditor On?*

The report issued by the Auditor has to be honest and factual. However, reports of this type are usually based on the Auditor's **subjective** assessment. Even where the Auditor's team carry out an **objective** assessment, the results have to be summarized, interpreted and analysed by Engineers who have different opinions and vastly different experiences.

There is therefore no practical reason why the Seller's Auditor cannot describe an asset as being in 'Excellent condition', even if the Buyer's Auditor regards it as 'Fair'. Neither term is precise or defined. The condition of the asset is purely a matter of opinion.

This is not devious or underhand. Potential Buyers expect information provided by the Seller to push the positives and gloss over the negative aspects of the operation. This is exactly why they employ their own Auditors to carry out additional reviews.

Typical promotional material is intended to sell a product or a service. We have all seen TV and media advertising so know full well the tricks advertisers use to get their message across. Auditors don't need to resort to smoke and mirrors when writing reports.

All they have to do is tell the truth (or at least, their version of the truth).

## *From the Seller's perspective*

The **Seller** wants the Auditor's team to highlight all the positive features of the scheme, minimize any potential downsides and put residual risks into perspective. The **Seller** does NOT want to highlight negative issues or anything that may put off potential investors or lenders.

## *From the Buyer's perspective*

The **Buyer** wants the Auditor's team to highlight potential technical problems, capacity constraints, or significant omissions in data or documentation. They want worst-case predictions on productivity highlighted, as well as potential construction liabilities and workface clashes identified.

The **Buyer** wants the Auditor's team to see the glass half empty. Not so empty that the Buyer walks away; just empty enough to be able to drive a harder bargain with the Seller.

## Stages of an Audit

A typical Audit will have the following stages, usually in this order;

- Clarify the role of the Auditor and the scope of the audit;
- Appoint the Engineers and Specialists to work on the Audit;
- Document Review (see below);
  - ❖ Identify what documents are required;
  - ❖ Review documents submitted
  - ❖ Identify shortfalls and additional information required;
  - ❖ Request missing information;
- Site visit (see below), including soliciting for missing documents, requesting new documents, asset condition review, and presentations by Local Management;
- Final review of documentation;

- Prepare audit report;
- Submit draft for comment;
- The final version of the report issued to the Client;

## Document Review

The majority of most audits are carried out in the comfort of the Auditor's home office. This is because a large proportion of the audit is based on documents supplied by the Terminal Owner.

In the case of a tank farm or terminal being bought or sold, the Auditor preparing a contribution to the Due Diligence process is only one of many organisations that require access to similar documents. Nowadays, it is common for an electronic Data Room to be set up and all documents viewed on line.

Most of the documents in the Data Room will have been existing documents, not prepared especially for the purpose. Frequently these documents will be out of date or not fully representative of the current situation. In most cases this will not be intentional but in some instances misleading data is presented intentionally. So while the Auditor's team is forced to use it, a degree of skepticism should be maintained. If in doubt, the Auditor should request clarification in writing prior to or during the site visit.

The biggest issue is frequently not what has been supplied but what is missing. If not included in the Data Room initially, the following information should be requested before the site meeting. This list deals specifically with the tanks and looks formidable. However, any competent Terminal Owner should have this data readily to hand, probably in a single report in the form of an Asset Register;

| | |
|---|---|
| Tank reference | O/fill protection |
| Tank Pit | Fire-internal protection |
| Tank type | Fire-external protection |
| Date constructed | Fire-actuation |
| Design code | Product stored |
| Diameter/Height/Nominal capacity | Mixers-style/number |
| Certified capacity | Heating-style |
| Material of construction | Heating medium |
| Lining | Temp.-max |
| Style of roof | Tank insulation |
| Style of floating roof | Double bottomed |
| Roof drainage | In service/Out of service |
| Floor style | Date of last internal inspection |
| Floor construction | Date of next internal inspection |
| Level indication-local | Status-Certified/Not certified |

The following information about the tank bund/tank pits should also be requested by the Auditor;

| | |
|---|---|
| Tank Pit Designation | Approximate capacity |
| Construction date | Design standard |
| Bund style (Pit/Dyke/Wall) | Largest tank in Pit |

| Material of construction-walls | Largest Tank in Pit capacity |
| Material of construction-floor | Surface water drains |
| Approximate height | Tertiary containment provided |

Request that this data be supplied as a spreadsheet. If the asset register is provided on a spreadsheet, it is easy to produce graphs showing the range of tank capacities, tank ages, liquids stored, the mix of tank types in inventory, etc.

In addition, the Auditors will be looking for information about how the business is running now, and how it has performed historically. This is considered in more detail-see **Reviewing Past Performance** below.

## Site Visits

Site visits are the only real opportunity for the Auditor to understand how the business operates. They allow the Auditor's teamto get a clearer understanding of the facilities and to meet the local Management Team for one-to-one briefings.

The objectives of a visit are;

- Confirm to potential Investors and Lenders that the facilities match the description;
- To verify verbally that there are no previously undisclosed issues such as PR, HR, land ownership, and potential constraints and bottlenecks;
- To interview local staff to get an understanding of how the site is managed;
- To verbally expedite documents essential for the document review but not received;
- To inspect the facilities to determine their condition;

### *Preparation*

Photographs taken during the site visit will help the Engineers to write their report. 'Due Diligence' reports can be dry and monotonous to read and any illustrations or diagrams that are included with the text will be appreciated by a non-technical audience. Permission to take the photographs should be requested well in advance of the site visit.

It is not uncommon for Terminals to resist: excuses are usually based on claims for safety and security. To counter these the Auditor should request that the Operator escorting them around take the photographs OR that the Terminal could provide an IS rated camera OR that digital copies of specific photos should be made available from the Terminal's photo library (yes, they all have one, even if they don't admit it).

If the Terminal say they will have to send the photographs after the visit, this is not a problem. It gives the Auditors the excuse to contact the Terminal directly to chase any other outstanding documents!

For an Auditor to go to a site without PPE is unforgivable. It is acceptable etiquette to ask to borrow a hard hat as everyone accepts they can be a pain to carry, especially if air travel is involved.If the Auditor doesn't take PPE, the best you can expect is a brief drive through and you will have only yourself to blame. To assume the Terminal will supply boots and coveralls during the visit is, frankly, insulting (although not unknown).

## Checklists

A detailed checklist is helpful to ensure that the audit covers all topics and that nothing gets left off. It also means that actions can be carried out in the most time-efficient and logical manner.

It is better to spend time in advance of the visit preparing the checklist, rather than trying to do something on the fly during the meeting. However, as the visit is an information-gathering exercise it is not feasible to plan too much in advance.

The Auditors should use checklists when having face-to-face meetings and during visits where time is precious. The checklists serve as an 'aide memoir' or a prompt. They are the starting point for dialogue and you should ask **supplementary questions**. The more you understand about the Owner's business, the more they tend to tell you. Very often the Terminal Owner and Terminal Operators will volunteer critical information that hadn't even been requested.

## Rules for Auditors

If there is a question, there should be a follow-up. Or a request for clarification on a specific point. Look them in the eye, be friendly (this is not an interrogation), and show interest. Be professional. Be complimentary when the reply warrants it. You are not there as their enemy. You are not there to trip anyone up.

When they finish answering the question, leave a pause at the end to see if they fill the gap with additional unrequested information.

A checklist is the starting point of discussions with the Owner's staff. It is only when talking to the site staff it will become apparent if some sections are irrelevant to that specific location or if additional topics need adding to the list.

## Reviewing the Asset's Condition

From the documents provided by the Owner, it is not possible to determine the actual condition and status of the facilities. The Investor/Lender may request the Auditor to assess the condition and status of the asset and report back. The principal concerns will be:

- Is the Owner's description of the asset fair and truthful?
- Are there obvious problems with the asset not already declared by the Owner?
- What is the physical condition of the asset?

Most Investors and Lenders will be interested to know the condition of the Terminal. However, the Employer's agreed scope of work <u>must</u> include this aspect and the level of detail required. It is not unknown for anEmployer, after receiving the draft report, to request that the condition of the assets be included. As this can only be achieved by reviewing all the latest tank inspection reports, it is impractical unless the requirement is built into the price.

However, the Client should be steered away from requesting this sort of detailed analysis. It takes a huge amount of time, is seldom conclusive and Clients don't understand the amount of work involved (or the potential costs-even if the Terminal can supply all the relevant documents).

## Information Sources

We must recognise the limitations of an assessment of the condition of the Terminal. Information can only be gained in a limited number of ways;

- Maintenance records provided by the Plant maintenance team;
- Anecdotal information provided by the Plant Maintenance team;
- Visual inspection of the facilities during a time or access limited visit to the site;

## Tank Condition

Routine out-of-service tank inspections, either to meet statutory requirements or as part of a RBI regime, confirm metal loss and residual wall thickness of the plates.

Although tanks are a high capital value asset, it is seldom that the Auditor can retrieve accurate data in a timely fashion. It should be possible for the Terminal Owner to provide a copy of the last tank inspection report. However, this was produced by the independent Tank Inspector to identify remedial work, not necessarily to quantify the extent of the remedial work. Therefore, whilst it is generally possible to identify where the defect (usually corrosion) was and its cause, it is not possible to see the extent of the remedial work. This would only be uncovered by checking the corresponding maintenance report, together with the final report from the Tank Inspector.

Even that does not indicate the potential service life for the tank, as corrosion could have been occurring over the last 20 years of the tank's life or it could have been only a recent event, caused by a change of operating conditions or contents.

We frequently see tanks up to 80 years old still in service, so it is no surprise the most frequent defect noted is corrosion. Tanks in hydrocarbon service typically have corrosion in the following areas;

- Conical tank roofs tend to buckle at the joint with the wind girder/shell plates. As a result, rainwater can 'pool' around the perimeter of the roof. Over time the constant wetting and drying of the roof will cause pinhole corrosion on the exterior of the roof plates, leading to perforation. This process is accelerated when the tanks are in a marine or coastal location.

- Even in hydrocarbon service, water accumulates inside the tank. As the tank is emptied and filled, and due to the changes in temperature between night and day, the tank 'breaths' and condensation can form on the inside. Rainwater from leaks in the water can enter the tank. Heating coils in the tank can leak, allowing steam condensate to enter the tank. Unprocessed hydrocarbon cargos can also have a quantity of BSW (bottoms, sludge, and water). The consequence is that two specific areas of the tank's interior can suffer accelerated corrosion. There is an area of the tank shell, typically all around the shell about 1 metre above the floor plates that corrodes. In addition, due to bacterial (anaerobic) corrosion, the interior surface of the floor plates can be corroded. This is a particular problem with the storage of jet fuels, where contamination of the stored product is unacceptable.

- The annular ring and the floor plates generally sit on a foundation (see below) and it is common for the water to penetrate by capillary action, causing corrosion on the exterior of the floor plates.

All of this corrosion is internal and can only be found when the tank is being inspected whilst out of service and empty.

For an Asset Condition Review, the only opportunity for the Consultant to assess the tank condition is by a site visit and the chance to talk to the maintenance personnel. Relying on anecdotal reports from the maintenance team may seem unwise, but they are usually a reliable source of information.

Many of the operating and maintenance staff have been on the site for a long time and have an unrivaled knowledge of the facilities. They are usually also very happy to share this information as they appreciate the opportunity to interact with Engineers from outside their own company, but who understand their problems (i.e. no time, no resources, and no money!).

In most cases, however, a 'walk by' during the site visit is the only chance for the Auditor to assess the condition of the tank, and this is limited to the condition of the exterior of the tank, as seen from the ground level.

### Tank Residual life

When the tank is returned to service following an internal examination, the residual life of the tank can be assumed to be the time before the report date and the next scheduled internal inspection. Defective parts of the tank can be replaced or repaired to extend the life of the tank almost indefinitely.

Repairs of any type and magnitude can be made to typical vertical steel storage tanks, including the complete replacement of the tank bottom by jacking the tank up. The main issue tends to be the size and suitability of the configuration of the tank for its future service, and whether the costs of the repairs are economically justified considering its current market value.

## Reviewing Past Performance

Depending on the Employer's concerns, the Auditor's review can include a variety of **Past Performance Audits**.

In most cases, the Employer has only a vague idea of what they want and looks to the Auditor for guidance. The Auditor should ensure that the Employer's requirements are clear before work starts. These additional (sub) audits may, if required, take place during the planned Site visit.

The Employer can 'mix and match' between the different reviews listed below. For instance, the Plant Performance Review can be based on KPI data which may be available in the VDR, so no action may be necessary during the site visit. OPEX/CAPEX data is difficult to obtain and may be unreliable so this is seldom requested by the Client.

These narrow, focused audits have to take place at the site as information is required from the relevant local management, operations staff, and maintenance staff. Once the site visit has taken place, it is not possible to go back for a second bite at the apple.

### Plant Performance Review

The objective is to provide proof, either graphically or in tabular form, that production rates demonstrate year-on-year improvements whilst downtime/costs, etc. reduce over time.

Different companies measure performance by different parameters. Some express it in volume throughput, others by weight or units produced. As a result, this review has to follow the Operator's normal method of recording. If the Owner uses KPIs as a management tool, the data required may be available in the VDR.

### Management Systems Review

The objective is to discover processes, systems, documentation, and controls in use. The review permits potential Investors the opportunity to see how the management systems are organized, and from this gauge the competency of the Owner's management team.

### Operations and Maintenance Review

The objective is to discover processes, systems, documentation, and controls in use. The review permits potential Investors the opportunity to see how the operations and maintenance systems are organized, and from this gauge the competency of the Owner's management team.

### Capex/Opex Review

A Capex/Opex review is a desktop study. It is normally performed by a Cost Engineer or Accountant and is not often requested by an Investor or Lender. However, it can provide valuable information on the financial performance of a business, its management team, and how it is likely to perform in the future. In particular, it may demonstrate the ability of the Owner's team to bring in projects on time and budget.

## C.3    Technical Audits

A technical audit (TA) is an external audit performed by a subject-matter expert to identify and evaluate defects, deficiencies, or potential areas of improvement in a process, system or proposal.

Technical audits can be carried out for a variety of reasons and are not exclusively associated with financial transactions and can include:

- Readiness Review;
- Pre-Commissioning Review;
- Technical Systems Review;
- Performance Evaluation;
- Supplier Audit;
- Surveillance (of Contractor or supplier);

Technical audits tend to be varied and unique for a particular situation. They can be as varied as monitoring organisational change, leading improvements to business systems and practices, and identifying problems with material contamination at a construction site.

### Objectives

To be fair, objective and balanced, the Technical Audit has these objectives:

- It must be a systematic and objective assessment:
- It should focus attention on the significant issues:
- It should eliminate trivial issues and help prioritize findings:

### Guidelines

The Technical Audit generally monitors activities against pre-agreed standards such as:

- Regulatory requirements;
- Company Standards;
- Company Procedures;
- Industry Standards;
- International Standards;
- National Standards;
- Engineering best practices;

The Auditor can identify problems but should not be responsible for developing specific solutions to those problems.  Only the Owner's management team has the authority to implement corrective action.

The Auditor should possess adequate professional competency to run the audit, including the relevant training, education and experience.

# APPENDIX D

# *New Terminal Design Review*

## D.1  Preface

One of the most common reasons for an Engineer to be asked to assess a tank farm or terminal is where the Owner plans to extend existing facilities (or build new facilities) and they need external financing from Investors and Lenders.

Developers looking for external finance for greenfield projects and startups may have limited construction experience and no track record of operating similar facilities, so Investors and Lenders are rightfully cautious.

In these cases, the Investors and Lenders frequently employ Engineering Consultancies to carry out independent reviews of the proposals. The actual work of the review is carried out by the Consultancy's employees or independent Sub-Contractors with specialist knowledge of the process or industry.

On major capital investment projects, it is normal to have at least two independent reviews. Typically, one Consultancy will be employed to consider **environmental issues** and another Consultancy will consider **technical issues**. Commercial and legal audits may also be carried out in parallel.

We will only consider the work of the technical review in the following pages.

## D.2  Terminology

To avoid any confusion, I am assuming there will be an **Engineer** who is a subject matter expert leading the team to review the Owner or Developer's proposals.

Existing terminals have **Owners**. Greenfield sites and startups are promoted by a new organization without an existing operating facility. This organization is the **Developer**.

When a Developer is trying to get Investors on board and Loans agreed, nobody moves past the point known as the **Financial Investment Decision** (FID). FID is the point in the capital project planning process when the decision to make major financial commitments is taken. Pre-FID, all proposals are unconfirmed and speculative. At the FID point, major equipment orders are placed, and contracts can be signed.

The work of the Engineer is to reflect on all technical aspects of the proposed development before any decision is made on FID. The issues the Engineer will consider are broken down into logical stages below.

When the proposed development is an extension to an existing tank farm or terminal, the existing facilities and the performance of the Owner's management team may be reviewed in a 'Due Diligence' audit (see **Appendix C.2**).

Where the development is a 'green field' or startup, the Engineer has to base their review solely on the Developer's plans and proposals. There is no past performance or existing asset condition with which to assess the Developer's competency to design and build a new facility.

The role of the Engineer is almost entirely subjective as they are considering plans and proposals that are speculative. The Engineer has to, based on prior experience with similar developments, identify potential technical, design and construction issues that may reduce the financial viability of the Developer's proposal.

## D.3 Design Review

The Engineer is appointed to review project feasibility and risks before FID, which can involve:

- Identification of technical concerns that could present financial risks for potential Investors;
- Put potential risks into context, to identify avoiding actions where feasible, and mitigating actions where the risk cannot be avoided;

Every situation is unique, so you can expect each design review to be different. To simplify the explanation, we can divide the issues into **Direct Risks** and **Indirect Risks**.

### Direct Risks

The information available for the Design Review will vary from project to project.

In some cases, the Developer will have incurred significant costs in preliminary development work before seeking the involvement of financial backers. In this case, a **Front End Engineering Design** (FEED) by a reputable design agency could already have been completed.

The philosophy of a FEED is to take an idea from Concept Development and do sufficient additional design engineering to reduce the technical and schedule risk (and consequential financial risk for the Developer). Standard practice would be to develop the design to the point where:

- A P&ID had been produced and basic safety studies have been carried out;
- Bills of Materials have been produced;
- Procurement packages for long lead items or materials have been completed;
- Construction Schedules have been produced;
- The project has been costed, using known equipment costs and Contractor estimates where possible;
- Piping, Civil, Electrical and Control & Instrumentation design have reached a point where Construction Contracts could be issued for Tenders, or an EPC contract could be signed with minimal risks for the parties involved;

In the event of all of this being available, the Engineer's job is relatively straightforward as it will only involve checking the work done by the design contractor. But to minimize the cost and schedule risk for the Lenders and Investors, the Developer will have already pre-invested significant sums.

If the Developer minimizes his pre-FID engineering expenditure, the financial proposition put to Lenders & Investors will have a lot more uncertainties attached, and the financial risk rises accordingly. If the Lenders & Investors proceed with the project, then we can presume it will be on more financially onerous terms or with caveats or penalties in the event of default by the Developer.

The overwhelming majority of schemes that come through the doors of the Engineering Consultancies are barely past the **Concept Development stage** (if that). Sketchy technical proposals are seldom well documented and assumptions are never spelled out. Engineering designs may be preliminary; costs may be based on historical data (and of uncertain accuracy), and schedules are invariably based on wishful thinking by people with limited experience.

It is not unknown for Banks to take the approach that it is the Engineer's job to sort out the mess, and if the Engineer has properly priced this into his bid, there is a reasonable logic to this. However, the Engineer cannot do the Developer's work and uncertainties will remain even when the Design Review has finished. The Consultant carrying out the design review cannot be expected to put the lipstick on a pig when they were hired just to give an objective and impartial opinion on someone else's project.

In the end, the costs come back to the Developer eventually, either through pre-FID engineering costs or increased post-FID financing costs.

It isn't an exaggeration to say that some Developers have a strange idea that a FEED is an avoidable expense. There have been plenty of cases where Developers have expected Construction Contractors to develop the Concept into a Detail Design (without schedule impact) for free. Needless to say, this is not a concept supported by the Construction Contractors.

In short, poorly developed schemes pose a particular problem for the Engineer and there should be a degree of caution used when pricing scopes of work to undertake any review.

Projects tend to fall mid-way between these extremes.

FEEDs are often of variable quality, if available at all. Developers often want to pass the work to a single **Engineering, Procurement and Construction** (EPC) contractor as this allows the engineering cost to be deferred until after FID. But it also means that few of the technical uncertainties are addressed before the Lenders/Investors commit to the project.

The Engineer is expected to do the majority of the assessment using data and documents supplied by the Developer. Sometimes the documentation is sketchy at best, and it is depressingly common for the Engineer to have to embark on a data collection mission just to collate the key facts.

The Engineer and his team should look at the following areas regardless of the quality of the documentation supplied;

## Permits

The Engineer will normally review the status of planning applications, environmental consents and construction permits. This is one of those areas of crossover as the Environmental Consultancy will carry out the same review, just from a different perspective.

Before FID, the best an Engineer can expect is a listing of all permits that will be required. At this early stage, it is unlikely permits will have been awarded. If Permits have been approved but contain constraints, restrictions, and caveats, the Engineer may request sight of the relevant correspondence between the parties concerned.

The Engineer should check:

- List of environmental permits, construction permits and operating licenses that will be needed before startup of the facilities,
- Status of those applications including date submitted;

- Copies of all approvals;
- Known problems, constraints or restrictions

## Technology

All new technology brings increased risk to Investors. The Engineer should check the scheme doesn't rely on novel construction methods, unproven technology, or technology that may raise legal restrictions due to copyright or Intellectual Property Rights or licensing constraints.

The Engineer should check:

- Use of untried technology;
- Use of technology with a limited successful track record;
- Novel or unusual use of current technology;
- Pushing boundaries with current technology;
- Licensed technology or processes;
- Anticipated IPR issues;
- Copyright issues;
- The volume of high-risk issues;

## Engineering

The Engineer needs to establish the extent of the engineering carried out by the Developer to date <u>and</u> the extent of the engineering to be done after FID. Engineering carried out after the FID increases the engineering risk, the risk of cost overruns and the schedule risk for Investors and Lenders.

The Engineer should check:

- The register of documents produced to date, as well as their status (final/draft), number of revisions and who has approved the issues;
- Copies of philosophy documents, especially project plan, contracting, project management and procurement;
- Status of the conceptual designs;
- Plot plans, floor plans, elevations, schematics for all ancillaries and services;
- Electrical single line diagrams;
- Instrument schematics;
- Who has done the engineering work to date and their prior experience;
- Lists of equipment (especially all safety-related and long lead time), equipment specifications, lists of proposed suppliers;
- Status of the planned construction contracts;

## Cost Estimate

The Developer should be asked for copies of ALL cost estimates produced to date, including explanations how the estimate was built up and sources of cost data.

In an ideal world, the Developer would have been out to an EPC contractor and received a fixed and firm offer. In this way, the final cost can be predicted with a reasonable degree of certainty.

If the Developer does not plan to use a single EPC contractor and also plans on procuring the long lead items themselves, the predicted costs may be reasonably close to the final cost.

At the other end of the range of possibilities will be the Developer who has a project facing unique challenges or site conditions. These estimates may be built up using equipment costs (from other old projects) and anticipated construction/labour costs etc. In this case, the historical data is frequently inaccurate. In addition, construction costs vary enormously from contract to contract and are difficult to compare on a like-for-like basis. If all the costs have not been fully identified, the contingencies have to increase, leading to an unreliable forecast cost.

These estimates are generally easy to identify, as they have many pages of line items and precise costs. However, whilst costs may be precise, they are seldom accurate. It is not unknown, in fact, it is fairly common, for final costs to be +40% of the original predicted cost at pre-FID stage.

The best cost estimate is usually for repetitive work, using a standardized design and where the Developer is using cost data from other recently completed schemes. Cost estimates derived in these cases have high reliability and have usually been prepared by competent Quantity Surveyors.

There are other ways of estimating costs. Some estimators have databases of piping costs per 100metre, and can estimate vessel costs based on the vessel weight and the current price of steel. Possibly the most misleading estimates are those produced by computer-based estimating packages. Whilst they superficially appear to be amazingly detailed (even to the point of including nuts and bolts) their historical database of costs is usually totally inadequate.

Unfortunately, we see too many Developers predict the final cost based on published costs (i.e. the typical cost of a terminal of this size, based on a Google search). These amount to little more than a wet finger in the air and these estimated costs should be treated with extreme caution unless the Developer can demonstrate they are comparing eggs with eggs.

Overestimating the accuracy of the cost estimates is frequent. Developers do not appreciate (or are reluctant to admit) that at the 'Concept' stage, they are unlikely to be better than +/-40%.

The Engineer should check:

- Who has produced the estimate, using what methodology;
- Claimed accuracy and supporting evidence;
- Sources of raw cost data, including material and equipment costs. Data last updated and by whom;
- Basis of calculating labour costs assumed productivity, etc.
- Was the equipment list and BoQ available to the estimator and how accurate;

**AACE RP 18R-97** gives detailed recommendations on the accuracy classes that can be used. Class 3 estimates are typically prepared to support full project funding requests, and become the first of the project phase control estimates against which all actual costs and resources will be monitored for variations in the budget. They are used as the project budget until replaced by more detailed estimates.

Typical accuracy ranges used by RP 18R for Class 3 estimates are (-10% to -20%)on the low side, and (+10% to +30%) on the high side, based on the assumption that 40% of the engineering development was complete when the estimate was prepared.

Based on the cost overruns being seen on major infrastructure projects worldwide, I would suggest these tolerances are overly optimistic. On the other hand, Developers would rather get a financial commitment from their Investors and Lenders early and then deal with the subsequent fallout when the costs start to escalate during construction. The two points nobody can dispute is that early warnings of escalating costs are always ignored, and nobody likes a Cassandra.

## Schedule

Schedules are frequently described as being between Level 1 and Level 5. Only the first three levels are relevant pre-FID:

- **Level 1** is a high-level schedule showing key milestones and summary activities only. Principally used as a graphical management decision-making aid. It is the Developer's prediction, not based on accurate or historical data.

- **Level 2** schedules are broken into smaller activities. Typically a Gantt (bar) chart format but rarely developed from a CPM network, assists in the project decision-making process. Used to identify project activities and deliverables.

- **Level 3** is usually at a level of detail to reflect all the contracting party interfaces. Seldom available pre-FID as contractor planning data is not usually available before FID.

When considering a new asset, the Developer has to put together a timeframe, as many of the costs they will incur are time-dependent. Whilst it would be nice if accurate and detailed information was available, before FID no construction contracts have been awarded so little or no information will have been supplied by contractors or suppliers.

Level 1 schedules frequently consist of less than 10 bars and no start or finish dates. The time axis generally starts at Month 0 and shows the predicted project duration in months (or years!). Another notable feature is that activities tend to overlap. These are the types of diagrams that are used in management reports or as illustrations in the Economic Model or Investment Proposal.

In short, they provide little information that the Consultancy can use for analysis and are little better than a crystal ball.

The Engineer should check:

- Level of most detailed schedule provided;
- Who produced the software package used andwhat skill level of the operator;
- Is Level 1 rolled up from Level 2 or a separate bar chart;
- Obvious omissions that could reduce overall duration;
- Is the network available for review?
- Are activities linked or free-floating (i.e. CPA or simple bar chart)?
- Level of activity considered-- is it detailed enough to draw conclusions?
- Calendars, holidays, seasonal constraints and work patterns reflected or ignored?
- Picking activities at random, are durations realistic? Are links and constraints realistic?
- Are equipment delivery durations realistic including shipping/customs etc.?
- Construction logic;

Level 2 schedules are a little better than Level 1, as they still don't include any contractor or supplier data such as activities, interfaces, or material lead times. They do at least show that the Developer has looked in more detail at the parts of the project and tried to anticipate activity durations and construction sequences.

Gantt charts with several hundred activities can provide a useful insight into the inner workings of the project for the auditor.

## Indirect Risks

These are many issues that could have an indirect impact on the project. In some ways, this can be regarded as a check on the quality of the work being done by the Developer's management team. The aim is to identify potential risks, options to eliminate the risk or mitigating actions.

The Engineer should check:

- Project Management Team. Developers usually provide details of how the project will be managed after FID (the Operator's Project Management Team):
- Where will the team be based;
- Organization of the Engineering phase;
- Organization of the Construction/Commissioning phase;
- Competencies and experience of personnel;
- Systems and Procedures (new or proven);
- Document control proposals;
- Commercial control and administration;
- Engineering Strategy;
- Operational Philosophy;
- End of Life Philosophy
- Construction Logic includes:
  - ❖ The sequence of activities to minimize interfaces and clashes;
  - ❖ Traffic management and security;
  - ❖ Maintaining quality and onsite inspection;
  - ❖ Safety;
  - ❖ Environmental issues;
  - ❖ Pre-commissioning activities and Commissioning;
  - ❖ Technology risks;

The Engineer needs to have a clear understanding of the Developer's progress to date.

···•••••●●●•••••••

# Appendix E

# *Tank Inspection Checklists*

## E.1    In Service Inspection

Most inspection regimes require an annual external inspection of the tank and foundation by a suitably qualified Tank Inspector. A full list of inspection items is contained in **EEMUA 159** but the significant items include the following:

### Shell and Roof

- Thickness measurement and remaining life calculation;
- Shell and roof plates inspection for corrosion and deformation;
- Shell and roof plates circumferential and longitudinal welding joints inspection for corrosion and cracking;
- Nozzles, man ways, flange welding joints inspection for corrosion and cracking;
- External part of annular plates inspection, specifically welding joint with shell plates;
- Side vents inspection;
- Gasket seating surface inspection of nozzles and man ways flange joints;
- Top angle, wind girder and stair way inspection for corrosion and deformation;
- Roof sump inspection (floating roof only);
- Pontoon and pontoon structure welding joints inspection for corrosion and cracking (floating roof only);
- Weather shield rubber and pontoon doors rubber inspection (floating roof only);
- Pontoon air leak testing (internal floating roof only);
- Painting or insulation inspection;

### Foundation

- Concrete foundation inspection for corrosion and cracking;
- Earthing system inspection;
- Asphalt inspection (around shell);
- Cathodic protection system inspection;

**✳ ✳ ✳**

## E.2    Out of Service Inspection

Most inspection regimes require a routine internal inspection of the tank by a suitably qualified Tank Inspector. A full list of inspection items is included in **EEMUA 159** but thesignificant items include the following in addition to the items listed for In-Service inspection:

### Shell

- Shell plate inspection for corrosion and deformation;
- Shell plate circumferential and longitudinal welding joints inspection for corrosion and cracking;
- Internal coating inspection (if any);

### Bottom Plate

- Thickness measurement and remaining life calculation on bottom plates and annular plates;
- Bottom plates and annular plates inspection for corrosion and deformation;
- Sump inspection;
- Internal coating inspection (if any);
- Vacuum testing inspection and supervision;

### Roof

- Roof plate inspection for corrosion and deformation;
- Roof structure member's inspection for corrosion and deformation;
- Internal coating inspection (if any);
- Internal Inspection-Floating Roof;
- Roof plate inspection for corrosion and deformation;
- Roof structure member's inspection for corrosion and deformation;
- above ground storage tank inspection;
- Roof plates and structure members welding joints inspection for corrosion and cracking;
- Roof sealing system inspection i.e. rubber seal, seal plate, etc.;
- Roof leg supports inspection;
- Pontoon gasoil leak testing;
- Internal coating inspection (if any);
- Roof drain system inspection;

### Equipment

- Gauge pole inspection for corrosion and deformation;
- Support and structure member's inspection for corrosion and deformation;
- Inspection of columns for corrosion and deformation;

· · · · · ● ● ● ● ● ● ● ● ● · ● · · · · · ·

# APPENDIX F

# *Process Design Drawings*

## F.1   Preface

Those of you who already work at a Terminal or Refinery (or in any other process-related industry) will be familiar with the typical design documents we use as everyday reference. However, those outside the industry may not have come across the two key design documents, the **Process Flow Diagram** and the **Process & Instrumentation Diagram**.

Symbols are a graphical representation of physical equipment installed on the field. There are a few ISO and British standards available that provide symbols and best practices to draw PFD and P&ID, such as, ISA S5.1, BS 5070, and ISO 10628.

## F.2   What is a PFD?

A Process Flow Diagram (PFD) is a simplified overview drawing that shows the relationships between major equipment in a process plant, using equipment symbols. You can visualize the flow of material within the plant with the help of these drawings.

A PFD provides a quick overview of the entire operating unit and shows mass, pressure and temperature changes, usually in tables around the edge of the drawing. The Flow Diagram is also used for visitor information and new employee training. It is one of the core documents for drawing the Plot Plant and P&ID.

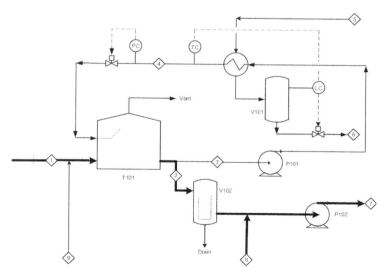

DIAG. E01 – Sample Process Flow Diagram

179

A typical PFD will include:

- All Major equipment: Each piece of equipment shown on PFD has a unique equipment number and a descriptive name. It also indicates the equipment's main dimensions, capacity, and operating information (this information is usually included in tables at the bottom of the sheet);
- The PFD follows the left-to-right approach for process flow. That means any process stream enters from the left and exits to the right;
- The process flow direction of all process lines;
- Control valves and process-critical valves;
- Major bypass and recirculation systems;
- Connections with other systems;

What is not included in a PFD?

- Pipe classes and pipeline numbers;
- Process control instruments;
- Minor bypass valves;
- Isolation and shutoff valves;
- Maintenance vents and drains;
- Relief valves and safety valves and
- Piping class

## F.3    What is a P&ID?

A Process & Instrumentation Diagram (P&ID) is a graphical representation of the process plant using various symbols that represent actual equipment.

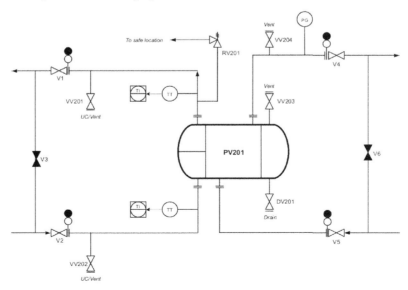

DIAG. E02 Process & Instrumentation Diagram

It is more complex than a PFD. If a system is shown on a single PFD, it will require multiple P&ID sheets to show the same system.

An important point to note is that what is shown on the P&ID is exactly what you get during construction. If the drawing shows a flange, there <u>will</u> be a flange. If the valve doesn't have a flange, it will be welded to the pipe. If the drawing shows a tee without flanges, then you <u>will</u> get a tee without flanges. Once approved for construction and being used as the basis of the safety reviews, the drawing <u>should not</u> be changed or altered without rerunning the safety reviews.

The P&ID is used for material procurement, construction, commissioning, cost estimating and construction planning. It is used to develop the plant layout, identify hazardous areas, and for preparing data sheets of equipment, valves, and instruments. The Construction Contractors use the P&ID to prepare bulk material take-offs for piping, electrical and instrumentation. It is a key document for safety reviews such as HAZOP, SIL, and operability.

During Operation, you have to maintain the P&ID in such a condition that it will show actual plant conditions at any time. It should be updated when any physical change is made so that the unit will remain compliant with codes, standards, and specifications, and can be operated safely under the defined process conditions.

P&IDs are used to train Operators and Engineers before they start work in the plant.

A typical P&ID will show:

- All the equipment, including installed spares and associated piping, including drain and vent line;
- Insulation or jacketing requirements;
- Instrumentation;
- Heat tracing and insulation detail;
- Information about utilities;
- Piping components, including their size, class, and tag number;
- Information required for design, construction, and operation, such as
    - A slope of the line;
    - Minimum and maximum distance from the equipment or instruments;
    - Minimum straight lengths after instruments;

•••••••••••••••••••••

# APPENDIX G

# *Acronyms & Abbreviations*

| | |
|---|---|
| AACE | Association for the Advancement of Cost Engineering |
| ADR | International treaty on the carriage of dangerous goods |
| AEO | Authorized Economic Operator |
| AFFF | Aqueous Film-Forming Foam |
| API | American Petroleum Institute |
| AST | Atmospheric Storage Tank |
| ARA | Amsterdam-Rotterdam-Antwerp chemical cluster |
| ASTM | American Society for Testing and Materials |
| ATEX | Equipment for Use in Explosive Atmospheres |
| BS | British Standard |
| BSW | Bottoms. Sludge & Water |
| CAR | Corrective Action Report |
| CETANE | Diesel equivalent to RON/MON |
| CIRIA | Construction Research and Information Association |
| COMAH | Control of Major Accident Hazards (UK legislation) |
| DEP | Design & Engineering Practice (Shell standards) |
| DCS | Distributed Control System |
| DWT | Deadweight (ships) |
| EEMUA | Engineering Equipment and Materials Users Association |
| EFR | External Floating Roof |
| EOV | Electrically Operated valve |
| ESD | Emergency Shut Down |
| EU | European Union |
| EV | Electric Vehicle |
| FAME | Fatty Acid Methyl Ester-Sustainable Diesel Substitute |
| FID | Financial Investment Decision |
| FSV | Fire Safety Valve |
| GA | General Aviation (non-commercial) |
| GHG | Green House Gases |
| HDD | Horizontal directional drilling |
| HGV | Heavy Goods Vehicle |
| IATA | International Air Transport Association |
| IBC | Intermediate Bulk Container |

| | |
|---|---|
| IFR | Internal Floating Roof |
| ILI | In Line Inspection (pipelines) |
| IMO | International Maritime Organization |
| IOC | International Oil Companies |
| IS | Intrinsic Safety |
| ISGOTT | International Safety Guide for Oil Tankers and Terminals |
| ISPS | International Ship & Port Security |
| JIG | Joint Inspection Group |
| kPa | Kilo Pascal |
| LEL | Lower Explosive Limit |
| LNG | Liquefied Natural Gas |
| LPG | Liquefied Petroleum Gas |
| M&A | Mergers and Acquisitions |
| MARPOL | International Convention for the Prevention of Pollution from Ships |
| MBM | Multi-Buoy Mooring |
| MCU | Motor Control Unit |
| MON | Motor Octane Number |
| MTBE | Octane enhancer |
| NFPA | National Fire Protection Association |
| OCIMF | Oil Companies International Marine Forum |
| OIML | international standard-setting body for metering |
| PCB | Polychlorinated Biphenyls |
| PLC | Programmable Logic Controller |
| PLEM | Pipeline End Manifold |
| PRV | Pressure Relief Valve |
| PTFE | Polytetrafluoroethylene |
| RBI | Risk Based Inspection |
| ROCE | Return on Capital Employed |
| RON | Research Octane Number |
| RoW | Right of Way |
| RVP | Reid Vapour Pressure |
| SCADA | Supervisory Control and Data Acquisition |
| SOLAS | International Convention for the Safety of Life at Sea |
| SPM | Single Point Mooring |
| SSSI | Site of Special Scientific Interest |
| TA | Technical Audit |
| TGS | Tank Gauging System |
| TMS | Terminal Management System |
| TVP | True Vapour Pressure |
| ULCC | Ultra Large Crude Carriers |

| UPS | Uninterruptable Power Supply |
| UVCE | Unconfined Vapour Cloud Explosion |
| VDR | Vendor's Data Room |
| VLCC | Very Large Crude Carriers |
| VOC | Volatile organic compound |
| VRV | Vacuum Relief Valve |

Printed in Great Britain
by Amazon

29020754R00117